PROTEIN CRYSTALLIZATION STRATEGIES FOR STRUCTURAL GENOMICS

IUL Biotechnology Series

***Igor F. Tsigelny**, Series Editor*

1. ***Protein Crystallization: Techniques, Strategies, and Tips.***
   A Laboratory Manual.
   Terese M. Bergfors, ed.
2. ***Pharmacophore Perception, Development, and Use in Drug Design.***
   Osman Güner, ed.
3. ***Protein Structure Prediction: Bioinformatic Approach.***
   Igor F. Tsigelny, ed.
4. ***Methods and Results in Crystallization of Membrane Proteins.***
   So Iwata, ed.
5. ***A–Z of Quantitative PCR.***
   Stephen A. Bustin, ed.
6. ***Comparative Enzymology of Cholinesterases.***
   Serge N. Moralev, Eugene V. Rozengart.
7. ***Protein Crystallization Strategies for Structural Genomics.***
   Naomi E. Chayen, ed.

PROTEIN CRYSTALLIZATION STRATEGIES FOR STRUCTURAL GENOMICS

edited by

NAOMI E. CHAYEN

INTERNATIONAL UNIVERSITY LINE
La Jolla, California

Library of Congress Cataloging-in-Publication Data

Protein Crystallization Strategies for Structural Genomics / edited by Naomi E. Chayen

p. ; cm. -- (IUL biotechnology series; 7)
Includes bibliographical references and index.
ISBN-13: 978-0-9720774-3-9 (hardcover : alk. paper)
ISBN-10: 0-9720774-3-X (hardcover : alk. paper)
1. Protein—Separation. 2. Crystallization. 3. Genomics. I. Chayen, Naomi E. II. International University Line. III. Series.
[DNLM: 1. Crystallization—methods. 2. Proteins—chemistry.
3. Computational Biology—methods. 4. Genetic Structures. 5. Genomics—methods.
6. Molecular Structure. QU 55.9 P9662 2007]

QP551.P695817 2007
572'.6—dc22

2007013270
CIP

Art work by Nina Bogoutcharoff.

Post Office Box 2525,
La Jolla, CA 92038-2525, USA

Library of Congress Catalog Card Number 2007013270

Printed in the United States of America

10 9 8 7 6 5 4 3 2 1

ISBN-10 0-9720774-3-X
ISBN-13 978-0-9720774-3-9 Hardcover

Contents

Contributors

James M. Berger
University of California, Berkeley
Berkeley, California, USA
jmberger@calmail.berkeley.edu
http://mcb.berkeley.edu/labs/berger/

Terese Bergfors
University of Uppsala
Uppsala Sweden
terese@xray.bmc.uu.se
http://xray.bmc.uu.se/terese

Martin Caffrey
University of Limerick
Limerick, Ireland
martin.caffrey@ul.ie
http://www.ul.ie/~ces/AcademicProfiles/MartinCaffrey.html

Naomi Chayen
Imperial College London,
London, United Kingdom
n.chayen@imperial.ac.uk
http://www.iobcr.org

Vadim Cherezov
The Scripps Research Institute
La Jolla, California, USA
vcherezo@scripps.edu

Charles G. Edmonds
National Institute of General Medical Sciences
Bethesda, Maryland, USA
edmondsc@nigms.nih.gov

Joseph C. Fanguy
Mississippi State University
Mississippi State, Mississippi, USA
jfanguy@otc.msstate.edu

Juan Manuel Garcia Ruiz
Instituto Andaluz de Ciencias de la Tierra
Granada, Spain
imgruiz@ugr.es

Charles S. Henry
Colorado State University
Fort Collins, Colorado, USA
cshenry@lamar.colostate.edu
http://www.chem.colostate.edu/cshenry/

Steven C. Holman
Mississippi State University
Mississippi State, Mississippi, USA
sch2@chemistry.msstate.edu
http://www.msstate.edu/dept/chemistry/rsch-sh.htm

Joseph D. Ng
University of Alabama in Huntsville
Huntsville, Alabama, USA
ngj@email.uah.edu
http://www.uah.edu/colleges/science/biology/joeng.htm

John C. Norvell
National Institute of General Medical Sciences
Bethesda, Maryland, USA
norvellj@nigms.nih.gov

Bernhard Rupp
q.e.d. life science discoveries
Livermore, California, USA
br@qedlife.com
bernhardrupp@sbcglobal.net
http://ruppweb.dyndns.org/

Joseph Valente
Novartis Pharmaceutical, Corp.
East Hanover, New Jersey, USA
joseph.valente@novartis.com

Julie Wilson
University of York
Heslington, York, United Kingdom
julie@ysbl.york.ac.uk

W. William Wilson
Mississippi State University
Mississippi State, Mississippi, USA
bilwil@chemistry.msstate.edu

Preface

We are currently living in an exciting age, where for the first time ever, human diseases are being understood at a molecular level. Protein crystallography plays a major role in this understanding because proteins, being the major machinery of living things, are often the targets for drugs. The fuction of these proteins is determined by their three-dimensional structures hence a detailed understanding of protein structure is essential for rational design of therapeutic treatments. Structural genomics, or more accurately, structural proteomics, which aims to determine the structures of thousands of proteins has emerged as a direct consequence of the genome project in which the human genes and other genes have been sequenced. Structural Genomics projects worldwide are detailed in http://www.isgo.org/.

Getting from gene to structure involves many steps: cloning, expression, solubilization, purification, crystallization, and only then, the determination of the structure. Once protein is pure and soluble, the key to crystallography is the availability of high-quality crystals. Producing high-quality crystals has always been the bottleneck to structure determination and with the advent of proteomics this problem is becoming increasingly acute. As a result, crystallization is gathering a new momentum as evidenced by the increasing numbers of commercial companies selling crystallization kits and tools, the increased investment of pharmaceutical companies in crystallization equipment and expertise, a high demand for practical courses in crystal growth, the launch by the International Union of Crystallography of a new journal, *Acta Crystallographica F*, formed in 2005 for publishing the recent explosion of data-concerning crystallization, and the establishment of the International Organization for Biological Crystallizaion (http://www.iobcr.org).

The past five years have seen some of the greatest achievements in the field of protein crystallization. It is now feasible to screen thousands of potential crystallization conditions by dispensing trials consisting of nanoliter volumes in a high-throughput mode. This has cut the time of setting up experiments from weeks to minutes, a scenario that was unimaginable a few years ago. Even more incredible, is the revelation that diffracting crystals can be produced from protein samples in volumes as small as 5–20 nanoliter. The subsequent phase of image capture and analysis of the crystallization drops is also progressing in great strides.

Surprisingly, in spite of the impressive advances accomplished, the crystallization problem has not been solved. High throughput has not yet resulted in high output and the current challenge is to design new and improved techniques (of screening and optimization) for the production of useful crystals. Scientists worldwide have taken on the challenge by tackling the crystallization problem from a variety of different aspects.

Research advances in recent years have opened up the scope for the development of new methods and tools to overcome the bottleneck of protein crystallization. A variety of parameters that could previously not be explored are now accessible thanks to sophisticated apparatus and the development of new science-based techniques to monitor and control the process of crystallization. However, in order to become useful to the structural genomics effort, it is vital to miniaturize and automate these techniques and adapt them to cope with the vast numbers of "leads" resulting from the high-throughput screening procedures. Such efforts are those of the immediate future and the focus of this book.

This book's aim is to assemble a selection of reviews highlighting the state of the art in techniques and strategies developed over the last four years for crystallization in the context of structural genomics. It is dedicated to the steps from the time of having pure protein to the crystallization, detection, imaging, and production of a diffracting crystal.

Most researchers involved in HTP consortia are engaged in solving structures and do not have the time or inclination to develop new methodology for crystallization. I have therefore chosen most of the Authors form the crystal growth community, for their expertise in development of methodology in their respective fields within crystallization.

The **first chapter** provides a background to structural genomics describing how it was conceived, funded, and implemented in the USA. This is just one example of many excellent Structural Genomics projects taking place worldwide (see http://www.isgo.org/) that could compile a book in their own right.

Chapter 2 provides an overview of setting up high-throughput structural proteomics projects in different environments and guidance as to how to handle the data and know-how resulting from the numerous experiments.

Chapters 3-5 describe different ways of conducting high-throughput initial screening for obtaining the first leads for crystallization. **Chapter 3** reviews the current robotic systems for screening that are available on the market for conducting high-throughput crystallization trials. Until 2002, screening for initial crystallization conditions was performed almost exclusively using vapor diffusion and microbatch. Since then, screening by the free interface and counter-diffusion methods that were least used in the past due to handling difficulties and the requirement for large quantities of sample, have been miniaturized and automated thus gaining a new lease of life. **Chapters 4** and **5** deal with miniaturization and adaptation to high throughput utilizing free interface and counter diffusion and gels, the former by microfluidics and the latter, using capillaries.

Chapter 6 describes a procedure for screening with a built-in optimization. Screening is performed using the standard screening kits but instead of looking for leads such as crystals, precipitation etc, the aim is to find the conditions that are at the metastable zone, which contains the ideal conditions for crystal growth.

Chapter 7 deals with strategies to optimize and improve the crystal quality once a lead has been obtained. These are methods that have been performed mainly manually and have recently been automated in order to adapt them to high-throughput experiments.

Chapter 8 describes the miniaturization and automation of experiments with membrane proteins in liquid cubic phase.

Dispensing trials with such amazing speed and ease has raised the issue of observation and monitoring of the vast numbers of trials. Major effort is currently being invested in designing image processing equipment for automated follow-up and analysis of the results as described in **Chapter 9**. Full automation of the visualization and monitoring of trials is expected to be the next major breakthrough in the field of crystallization.

The success rate of obtaining high-quality crystals is improving and the coming years promise to bring further advances in the more complex techniques that will play a major role in crystallization and proteomics. This will raise the rate of producing high-quality crystals, and will equip the genome project to deal with its awesome task.

Naomi Chayen
London, England, May 2007

1

Structural Genomics in the United States: The NIGMS Protein Structure Initiative

Charles G. Edmonds and John C. Norvell

1

Structural Genomics in the United States: The NIGMS Protein Structure Initiative

Charles G. Edmonds and John C. Norvell

National Institute of General Medical Sciences
Bethesda, Maryland, USA

Building on the success of the sequencing of the genomes of human and other organisms, numerous large projects have been undertaken for the high-throughput characterization of the proteins, one of the major products of these large numbers of genes. The new field of structural genomics aims at large-scale structure determination of a collection of proteins, an important step in inferring their function. This emerging field is made possible by new and more powerful crystallographic and NMR methods and instruments. The choice of the term, genomics, in the name is appropriate for this new, discovery-driven approach, since this effort is based fundamentally on the library of genome sequences.

Research in this field began in the late 1990s in a number of locations. As interest grew, the National Institute of General Medical Sciences (NIGMS) collaborated with Wellcome Trust (United Kingdom), RIKEN (Japan), and other international agencies to organize meetings beginning in 2000 in Cambridge, United Kingdom; Virginia, United States; and Yokohama, Japan, to discuss international coordination and to address the policy issues associated with the several structural genomics projects. At these meetings, scientists and agency officials formulated an agreement of principles and the establishment of a scientific organization for this field, the International Structural Genomics Organization (ISGO, http://www.isgo.org). ISGO International conferences on structural genomics take place bi-annually, with the last conference taken place in October 2006 in Beijing, China.

In the United States, early support for structural genomics came through research project support from both the NIGMS and the Office of Biological and Environmental Research of the U.S. Department of Energy. In 1999, NIGMS organized a national structural genomics program, the Protein Structure Initiative (PSI, http://www.nigms.nih.gov/psi). The long range goal of the PSI is to make the three-dimensional atomic level structures of most proteins available from their DNA sequences alone. This is to be accomplished by high-throughput execution of all steps beginning with the selection of the gene sequence and leading to the determination of the structure of representatives of sequence families. The PSI, like most structural genomics projects, emphasizes the creation of a catalogue of representative structures that will serve as templates for computational modeling of the structures of the larger set of protein family members. The availability of these experimental and computationally determined structures is expected to aid researchers in developing hypotheses and designing experiments, to have impact on a range of biological research problems, and to aid the development of new therapeutics.

The NIGMS Protein Structure Initiative was formally initiated in September of 2000 with the funding of seven research centers, each charged with the development of a pipeline for the high-throughput production and structure determination of protein targets identified as unique and representatives of protein sequence families. These centers, together with two additional centers, which were funded one year later, were combined with a number of program projects, individual investigator research projects, and small business innovation research and small business technology transfer projects to constitute a five-year pilot phase of the PSI. The aims of this initial phase were to test strategies and approaches, to build the structural pipeline, to develop new methods and instruments needed to lower costs, to increase success rates, and to automate all steps of the production and structural determination of proteins.

The nine PSI pilot centers that were funded by NIGMS are listed in http://www.nigms.nih.gov/Initiatives/PSI/Background/PilotFacts.htm.

Two of these centers, the Structural Genomics of Pathogenic Protozoa Consortium and the TB Structural Genomics Consortium were cofunded by the NIH National Institute of Allergy and Infectious Diseases.

Most of these pilot centers have chosen unique protein targets that have no close relatives with known structures in the Protein Data Bank (PDB), i. e., having no more than 30% sequence identity. Major developments in technology and methods have led to improvements in every stage of the structure determination process. Parallel processing of multiple protein targets and the high-throughput mode of operation have led to significant efficiencies and improved success for the determination of the structure of family representatives. Development of automation, standard operating procedures, laboratory information management systems, and factory-like decision tree operation has increased the throughput. More details on the pilot phase progress can be found on the PSI webpage given above.

During the five-year pilot phase, which is now nearing completion, the PSI has sponsored numerous workshops to aid in the planning of this large research program. In addition, technical workshops were held by participants in the initiative to address technical difficulties slowing high-throughput operation and barriers to high success rates. These meetings have been focused on topics such as target selection, data management, homology modeling, and NMR methods. Among these the most important and popular were annual "bottlenecks" meetings during which active laboratory participants in the initiative exchange their first-hand experiences regarding issues which reduce efficiency, economy, success rate, and throughput.

During the first eighteen months of the pilot effort, the initiative participants began to address, individually and collectively, the technical problems arising from the constituent tasks of structural genomics. Progress in such a multifaceted process was inevitably uneven. Barriers defined principally by the operational and engineering characteristics of the problem were more readily overcome. The obvious emphasis on robotics and automation, parallelization, and miniaturization rapidly yielded results in areas such as the automation of the X-ray crystallography beamline operations, the automated determination of protein structure from X-ray diffraction or nuclear magnetic resonance experimental data, and the large-scale management of laboratory information. However, against this background there emerged several more durable bottlenecks where barriers were more deeply imbedded in the complexities of cell and molecular biology, DNA and protein structure.

In early 2002, PSI participants convened a bottlenecks workshop (Protein Production Workshop, http://www.nigms.nih.gov/news/reports/protein_production.html), which focused on the problems high-throughput methods for cloning, production, and purification of proteins suitable for X-ray crystallography and NMR studies. Overall, the best results were obtained using parallel strategies which provide multiple paths to a successful outcome for the cloning, expression, and purification of a particular target.

A year later in the spring of 2003, a second bottlenecks workshop on this topic was convened (Protein Production and Crystallization Workshop, http://www.nigms.nih.gov/psi/meetings/protein_production.html). The scope for the agenda was expanded to include issues of protein crystallization. The workshop reflected a better understanding of the field and described numerous projects underway without offering dramatic breakthroughs. Substantial challenges remained to be addressed.

Three further Protein Production and Crystallization workshops have been held in 2004, 2005, and most recently in 2006.

The goal of all of these meetings has been to offer scientists a platform to share and exchange ideas and data, to discuss progress and problems, and to address current challenges. The result has been a multiplication of contacts and collaborations among the participating groups.

The accumulation of data on effective protein production and crystallization for many genes and proteins is having an important impact on the Protein Structure Initiative and on the field of structural biology as a whole. Data on both the positive and the negative outcomes of experiments on all structural genomics targets are valuable and are compiled in the Protein Expression, Purification and Crystallization Database (PEPCDB, http://pepcdb.rutgers.edu/). This database is available for reference and data mining to all researchers, giving information on target identity, pipeline status history, pipeline stop work conditions, and details of protocols employed.

When elements of the pipeline are well understood and well controlled, the high-throughput approach offers progress toward structures in the most rapid, efficient, and cost-effective way possible. When these elements are not fully understood or controlled, high throughput permits a rapid, economical, and efficient screening approach in which multiple paths are traced, hopefully to at least one satisfactory outcome. The abundance of representative family targets in the multiple sequenced genomes is a particular advantage in this respect.

The pilot phase of the PSI involved major investments in technology and methodology for making each step of experimental structure determination more efficient, less expensive, and more likely to succeed. During this first phase, which ended in the autumn of 2005, we have learned that structural genomics pipelines can be constructed and scaled up. High-throughput operation works for many proteins and has led to more than one thousand successful PSI protein structure determinations. However, as expected some protein classes are not yet amenable to this approach. These classes include membrane proteins, proteins from human and other higher eukaryotes, and small protein complexes. It is clear that the structural genomics approach of identifying multiple representatives of protein sequence families as targets can be successful. The development and refinement of methods and techniques continues. Where initial trials are not successful, the pilot centers have developed "salvage" procedures taking alternative routes for proteins that have failed at one of the standard steps of the pipeline.

The second phase of PSI, sponsored by NIGMS and the National Center for Research Resources (NCRR) started in July 2005. The focus has shifted to a production phase during which 10 new centers will use methods developed during the initial pilot period to rapidly determine thousands of protein structures. The PSI production phase (PSI-2) includes two types of centers: large-scale and specialized centers.

The large-scale centers, all established during the PSI pilot phase, have constructed structural genomics pipelines for the production and structural determination of proteins in a high-throughput operation. Over the five-year period of PSI-2, these centers are expected to coordinate their efforts to generate more than 4,000 structures of protein families, and thus meet the program's goals of increasing structural coverage of sequenced genes and providing benefits for the entire biomedical community. The centers will also develop new technologies to improve the pipeline and to increase success rates and lower costs. Each center also has an individual structural genomics research project.

The specialized centers will develop innovative methods, approaches, and technologies for producing and determining the structures of proteins that traditionally have been difficult to study. These include small protein complexes, membrane proteins, and proteins from higher organisms, including humans. During the next five years, these efforts are expected to overcome bottlenecks in the structural genomics pipeline, approach high-throughput operation, and determine a significant number of challenging protein structures.

Information on the centers and their activities can be found at http://www.nigms.nih.gov/Initiatives/PSI/Centers/

The PSI production phase will also facilitate the dissemination of the valuable data generated by the PSI centers through a variety of resources. These will include a materials repository for the distribution of clones generated by the centers; a Protein Expression, Production and Crystalliza-tion Database (PEPCDB), which captures the results of all the individual experiments undertaken by the centers; and a web-based Knowledge Base, which will act as a portal for the scientific community to access the products of the PSI. Network policies, including the inclusion of targets nominated by the scientific community and prompt data deposition and release, will ensure maximum benefit to the scientific community.

The spectacular wealth of information in the human and other organism genomes points to a logical line from gene sequence through structure to the large-scale understanding of the individual and collective function of entire proteomes. Homology modeling provides a shortcut for the daunting task of the individual structure determination of these numerous gene products. A systematic program for the determination of the structure of representatives of protein families will yield a large number of experimental structures and a larger number of computed structures. Each of these structures may serve to illuminate structure-function relationships and to facilitate the formulation of better hypotheses and the design of better experiments in structural biology. The resulting acceleration of both basic and applied biological discovery will have important benefits for research that aims to improve the diagnosis, treatment, and prevention of human disease.

2

A Guide to Automation and Data Handling in Protein Crystallization

Bernhard Rupp

Abstract

Substantial funding of public and commercial structural proteomics efforts has led to accelerated technical development and increased availability of high-throughput robotics for crystallography. Automation deployed in protein crystallography extends from parallel, small-scale protein expression screening to fully automated protein crystallization systems, and finally automated sample mounting, data collection, and structure solution and refinement. Successful implementation of laboratory automation requires careful planning with a balance of expectations and resources versus long-term costs and return on investment. It is important to recognize that not every research environment requires the same type or degree of automation, and operations research and process analysis are necessary to select the robotic automation layout most suitable to achieve maximal overall efficiency. Based on process outlines and a review of critical steps we discuss principles of cost-effective modular designs for crystallography-laboratory automation, emphasizing the importance of automated process scheduling, data capture, and adequate database support.

2

A Guide to Automation and Data Handling in Protein Crystallization

Bernhard Rupp

q.e.d. life science discoveries, Livermore, California, USA

2.1. Introduction

A significant influx of public and commercial funding over the recent years has contributed to rapid technical developments in automated high-throughput crystallization. On the public side, the NIH Protein Structure Initiative, PSI-I,[1] underway since fall of 2000, provided significant funding to nine P50 Structural Genomics (SG) Pilot Centers in the United States, and similar initiatives in Europe (SPINE), Canada, Japan, and Israel have been funded (the ISGO web site www.isgo.org/ provides a comprehensive up-to-date listing of all SG initiatives world wide). One of the main objectives of these centers is the advancement of high-throughput crystallography including methods for high-throughput protein crystallization.

Given the potentially enormous rewards of structure guided drug development,[2-4] it is no surprise that a substantial number of biotech ventures were able to attract capital to develop and implcment high-throughput techniques, in particular advanced robotic automation.[5,6] Closely following the pioneering technology development efforts, many vendors have now realized the market potential for automation of crystallization, and are competing with a wide palette of products ranging from basic liquid handling robots to integrated high-throughput crystallization workstations, imaging stations, and crystal-mounting robots. Since fall of 2006, the NIH funds four large scale PSI-II production centers as well as six smaller specialized efforts, and the development of sophisticated techniques and new robotics continues.

For the practitioner of crystallization, the choices of methods, instruments, and degree of automation can be bewildering. As none of the automation components or solutions offered are cheap (although good value for modest investment is available), the question arises how to spend available funds most wisely to achieve maximum return on the investment into robotics. Each specific situation might indeed require a different approach to implementation of automation technology. To guide in the assessment of a laboratory's need for automation, several key questions should be asked; for example: Is the research environment in which the deployment of automation takes place academically or industrially oriented? What are the expectations in terms of throughput? Which degree of automation is achievable given the financial resources? Which methods and experimental designs are best suitable given the particular environment and research objective? Will the information technology (IT) infrastructure be able to support and handle high-throughput data flow and process control?

This chapter attempts to provide some guidance to researchers interested in deployment of automation in their own laboratory. We will first investigate general aspects and objectives of high-throughput (HT) protein crystallization (and HT protein crystallography as a whole), followed by a brief recollection of the key principles in crystallization and their importance relative to automation and throughput. Finally, we will discuss practical implementation issues based on process analysis, and close with a brief synopsis of possible scenarios and examples for automation equipment. Detailed descriptions of current robotic systems are provided in a separate chapter.

It needs to be emphasized that this chapter is a *guide to deployment of automation* and not a "*buyer's guide*." Although principal advantages of certain procedures or types of equipment are discussed, the use of a specific example does not construe an endorsement for the product nor does the omission of a vendor or product mean that anything is wrong with it. Progress and innovation in the field of automation is swift, and web searches and updates will be necessary to discover the latest developments. We encourage discussion with experienced colleagues who have already paid their dues during their automation deployment—vendor specifications should be evaluated with some suspicion; of particular concern are issues of software interoperability and adaptability, which are often subtly neglected in flashy propaganda material.

2.2. High-Throughput Protein Structure Determination

2.2.1. Why High Throughput

What exactly is the motivation behind the desire to automate and to pursue high throughput in protein crystallography? To a large degree, the driving incentive behind structure determination on a "genomic scale" is the elucidation of the structural basis for molecular function—not just for one enzyme but for whole functional pathways. Such aspects will become even more important as the classic single gene—single disease targets become increasingly sparse[7] or new classes of pathogen-specific proteins become available as drug targets.[8] The goal of producing lead structures suitable for therapeutic drug design requires that such structures are of high quality and provide enough accurate detail for *in-silico* screening[9,10] or structure-guided lead optimization[2] Structures obtained for drug design purposes will naturally demand much higher criteria for quality compared to structures determined with the goal to populate the protein fold space, an explicit aim of the PSI-I projects.[1] In drug design, structures of multiple orthologs, numerous functional mutants, or multiple potential drug–protein complexes have to be determined. The need for high throughput, without compromise in structure quality, becomes evident, in particular to take the step from structure-guided lead optimization to true structure-based drug discovery.[11]

2.2.2. Definition of High Throughput

The meaning of the prefix "high throughput" in high throughput protein crystallography (HTPX) is rather diffuse and includes work ranging from modest scale—academic efforts focusing on affordable automation,[12,13] to truly industrial scale—commercially driven ventures processing hundreds of crystals of pharmaceutically relevant targets a week.[14-17] The underlying physical and methodical principles, however, are the same for HTPX as for conventional "Low-Throughput" (LT) protein crystallography. Most of the improved methods and techniques developed for high-throughput purposes are therefore also applicable—and indeed very useful—for LT efforts. Protein crystallography in general has made enormous technical progress through increased public and commercial funding devoted to the development of high-throughput technologies since the late 1990s.

2.2.3. High Throughput versus Low Throughput

The most significant difference separating "true" HTPX and conventional LT crystallography efforts is the amount of automation and process integration that is implemented. In contrast to readily affordable computational prowess, full robotic automation nearly always comes at a substantial cost, indicating the most significant rift between academic low- to medium-throughput laboratories and commercial HTPX ventures. Financial, infrastructural, and production-related constraints, for example, are very different in academic and commercial environments, as are efficiency considerations (**Table 2.1**): In a high-throughput environment it may be neither necessary nor efficient to pursue every recalcitrant target to completion, while in an academic setting careers can depend on the determination one specific structure.

2.2.4. Estimation of Throughput

It is instructional to review the success statistics from the NIH PSI-I initiative, which are available from http://targetdb.pdb.org/[18] in order to acquire a baseline for the attrition through a crystallography project from gene targeting through the major steps of a structure determination, up to the deposition of a validated structure in the PDB. Despite several caveats outlined later, the trend is quite sobering and shows that the attrition rate is substantial (**Fig. 2.1**). We must also recognize that the protein sample sets used by the PSI initiatives are strongly clustered. Most PSI efforts aim primarily for small prokaryotically expressed and high-

Table 2.1. Listing of key "environmental" factors in academic versus industrial research efforts and their impact on automation deployment.

Typical academic laboratory:	**HTPX drug discovery effort:**
Modest throughput requirements	Substantial throughput necessary
Highly differing proteins	Repeatedly same target proteins
High commitment to target	Ready to re-engineer targets, orthologs
Limited public funding	Substantial financial resources
Limited IT infrastructure	Highly organized IT department
Inexpensive graduate-student labor	Expensive full-time employees with industrial salary
Fluctuating funding	Relatively stable, venture capital, public stock offerings
Semi-automated process:	**Optimal automation level:**
Modular setup, manual transfers	Fully integrated robotic liquid handling and crystallization plate-setup robots
Liquid-handling robot	Integrated plate storage and image recognition and analysis system
Crystallization plate-setup robot	
Stand-alone image acquisition	Automated feedback to optimization setup, process scheduling
Manual plate storage, retrieval	Investment in novel automated crystal harvesting and robotic data collection on own synchrotron beam line
Manual crystal mounting	
Use robotic data collection at a synchrotron on demand	

ly soluble proteins ("low-hanging fruit") to meet throughput goals. Such proteins will likely have a different success space and higher success rate expectations than large complex assemblies or membrane proteins[19-21] or eukaryotically expressed proteins, in particular in the presence of conformationally heterogeneous posttranslational modifications and decorations.[22] Remarkable differences in success rates, even between prokaryotic organisms from which the protein target genes were derived, have already been reported.[17]

Even the best-case statistics (not always the best center in one category scores also best in another category) show that slightly more than one-third of soluble proteins form something identifiable as a crystal (the average number here seems to be more realistic than the best due to exaggerated metrics as to what constitutes a crystal); and that about

Statistic	PSI average (%)		PSI Best (%)	
Selected targets to *clones*	49.0	(49.9)	95.6	(95.6)
Clones that *express*	67.8	(33.8)	88.4	(84.5)
Clones to *purified protein*	16.6	(5.6)	81.7	(69.0)
purified protein to *crystal*	37.3	(2.1)	81.9	(56.5)
purified protein to *data*	14.2	(0.8)	33.6	(19.0)
purified protein to *structure*	12.3	(0.7)	22.5	(12.7)

Figure 2.1. Success statistics of NIH protein structure initiative (PSI) centers compiled from http://targetdb.pdb.org/ as of first quarter of 2004. Success rates in percent given for the average center and the best center, respectively. Columns in parentheses indicate cumulative success rate. The statistics do not differentiate between centers pursuing "high-hanging fruit" (e.g., human targets) while some centers also entertain tight connections with commercial high-throughput structural genomics ventures. The reporting metrics also vary between centers, and the opinions what constitutes a "crystal" differ widely.

another 30–50% of those crystals actually yield data allowing structure solution. In the absence of any efforts in protein engineering , we thus have no more than a 10–20% chance on average of obtaining a structure from a purified protein. The losses from construct to purified protein are equally dramatic (in particular, for other than soluble, garden variety bacterial proteins), and success rates are again in the order of about 20–30%. Clearly, protein production is crucial to success of a proteomics pipeline, as emphasized by the increasing importance of exploring multiple constructs, either from orthologs[23] or via protein engineering[24-26] to obtain inherently crystallizable proteins.

A process stage associated with at least 50% losses seems to be the step from harvestable crystal to diffraction data. There are multiple handling steps involved subsequent to observation of a crystal. Harvesting, cryoprotection, mounting and annealing,[27] radiation damage,[28] etc., are known to affect diffraction quality.[29] These parameters are introduced *after* the crystallization process and—provided they are screened in a suitable experimental design—do contribute to additional mineable data for the *total* process, using a combined success indicator "diffraction quality." Just as properties of each protein construct and the parameters of protein production influence the outcome of the crystallization experiment and thus need to be recorded for data mining purposes, the post-crystallization handling can provide equally important insights informing about the overall process efficiency.

Absent from the statistics in **Fig. 2.1** is another interesting measure—the cost per structure determination. In the PSI-I initiatives, the cost per structure (S) decreased dramatically from $500K/S in 2000 to $150K/S in late 2003 and is expected to decrease further to about $50K/S in 2005 when the initial phase of the PSI will end (Source: NIH report, Planning for the Protein Structure Initiative, Sept 2003). To what degree this decrease in cost is due to increasing efficiency of the centers' operation in general or a result specific to the deployment of automation, is not entirely clear. In any case, the projected numbers are significantly below the average total cost of $150–240K/S estimated by the NIH for "conventional" structural biology efforts.

2.2.5. The Data Explosion

Although the objective of a structure determination project may vary widely depending on the research environment, a top-level flow analysis of the experimental progress in the proteomics laboratory shows that in principle we follow multiple levels of successively branched screening steps, traversing a decision tree specific for our process implementation (**Fig. 2.2**). Despite the considerable attrition experienced, an enormous amount of data is generated that rapidly can outstrip our capability to analyze them.[30] The need for a suitable IT infrastructure to properly handle the data and to allow data mining for knowledge discovery becomes evident. Particular attention should be paid to full capture of all information available regarding the proteins to be crystallized. Often even what is known is not or only inconsistently recorded in the crystallization databases (or in obscure spreadsheets). Proteins in solution are "*sticky, dirty creatures that pick up no end of detritus from their environment*",[31] and protein preparations contain numerous reagents and additives such as lipids, detergents, or cofactors acquired throughout the course of purification, often only discovered once the structure is determined.[32-35] Each of these components may play a critical role in the crystallization process. For the success of classification-based machine-learning techniques, availability and quality of protein and purification related information is thus crucial.[36]

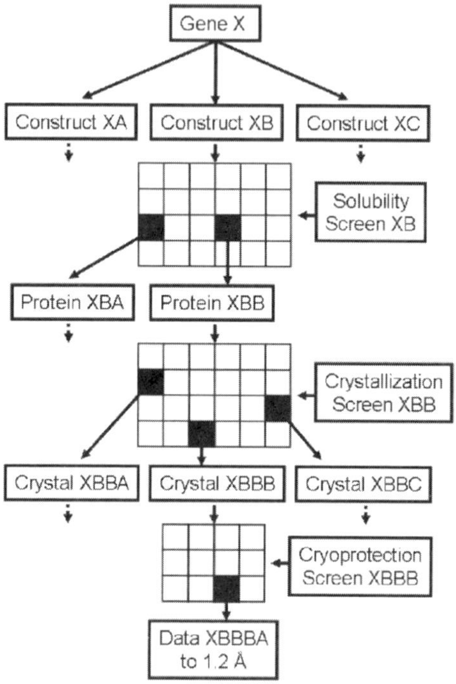

Figure 2.2. Path through a simple, hypothetical three-step process tree in a protein production and crystallization laboratory. In each step, invariant parameters from the input serve as classification basis for the outcomes of the associated screen. Despite substantial attrition, the amount of data multiplies dangerously[30] with each screening step. The complete set of information from gene to data can be contained as a single instance in case-based reasoning techniques.[37,38] Systematic cryoscreening of course requires that more that one crystal is available from the crystallization screening experiments. Figure reproduced from the work.[36]

2.2.6. Information Technology Infrastructure

Although it may seem initially excessive for a small effort, capture of primary data at the source, directly into a relational database via automated scripts or a laboratory information management system (LIMS), is highly recommended. Elaborate LIMS packages are mandatory for the pharmaceutical and biotech industry[39] largely due to regulatory requirements (21 CFR 11 compliance), but smaller and flexible systems are becoming available for proteomics and crystallography[40-42] and some are being developed by the PSI centers.[43] With increasing automation, tightly integrated process control, multiple feedback steps, real-time data processing and decision making, as well as machine learning for predictive purposes, are becoming a major component of any automation-intensive laboratory. Although initial cost and effort to implement a LIMS are not insignificant, the return on investment (ROI) can be less than a year.[44] The May-2004 issue of *Scientific Computing and Instrumentation* focuses on LIMS implementation and includes vendor information.

A solid relational database management system will form the core of any data repository, and must adequately reflect both the data structure and the process flow.[38] The database design should anticipate the kind of analysis and data mining to be performed[37], and a well designed data repository will support access to existing protein databases allowing cross-database queries of supporting information that can be used at various levels in the decision-making process.[45] At the top level, the need to construct new ontologies for data description and data mining of complex knowledge[46] needs to be anticipated.

The data amassed during automated protein production, crystallization, structure solution, and validation are of very heterogeneous nature and play an important role in different aspects of process analysis. Unfortunately, at this point it may already be decided whether our database is copious with hidden information and ready to be data mined for intelligent decision making support, or whether our production and crystallization data are presented in a disastrous agglomeration of heterogeneous sources. Distributed efforts or consortia are particularly subject to greatly differing and heterogeneous data repository objects. From the author's painful experience, the worst legacy types are perhaps spreadsheets distributed in undocumented network file systems.

2.2.7. Data Mining and Machine Learning

The purpose of crystallization data mining is to *establish relationships* between the experimental parameters and the experimental outcome. The relationships allow deriving *rules and predictions* for the outcome of a new experiment based on the *level of knowledge* we have gained by *prior analysis*. At the lowest level of crystallization data analysis, one might seek to make inferences about global "hot spots" in success space by basic frequency or propensity statistics, resembling the way the initial sparse matrix set[47] was conceived. At a higher level, one might attempt to categorize proteins into classes with distinct properties,[45] for example, to derive special screening kits for membrane proteins.[19] Ultimately, we would like to predict the outcome of a crystallization experiment for a specific protein construct of interest, given all the information we have available. In other words, important for the long run are correlations (generalizations) of the success distribution in crystallization parameter space with known properties of the protein, i.e., *known priors* that modify our choice of hypothesis (prediction) where to find the highest likelihood for success given what we know already—a classical Bayesian strategy.[36]

One might assume that the analysis of massive amounts of proteomics and crystallization trial data resulting from automated screening should readily allow deployment of statistical methods and machine learning to develop predictive algorithms for effective protein production and/or crystallization. It is not as straightforward as it appears. Independent of automation, the sources for the difficulties are founded in the complex physicochemical nature of protein production and crystallization, resulting in nontrivial experimental design issues, which again affect data consistency and data validity and thus, in turn, determine the applicability and significance of the data-mining algorithms that can be deployed to tackle the problem. It is important that under pressure to produce structures (the ultimate goal of most structural biology or drug-design laboratories), the opportunity to create comprehensive and consistent databases is not lost. This concern is not entirely unfounded as both the omission of negative results and the lack of the most basic quantity in statistics—the number of trials—have rendered publicly available crystallization databases (Biological Macromolecule Crystallization Database, BMCD;[48] Protein Data Bank, PDB)[49] virtually ineffective for the purpose of rigorous statistical analysis and machine learning, even with significant restructuring and annotation efforts.[50,51] An extensive review of data mining for predictive models for protein crystallization has been published recently.[36] Just turning loose some machine-learning algorithm on messy crystallization data and expecting to obtain meaningful patterns is wishful thinking.

2.3. Overview of Protein Crystallization

Informed decision regarding the impact of automation is greatly aided by a clear understanding of the key steps in the process and their dependencies and synergisms. A quite generic and basic scenario is outlined in **Fig. 2.3**. Generally, tasks in a protein crystallization laboratory will include obvious steps such as basic liquid handling, preparation of crystallization cocktails, crystallization plate setup, automated crystal detection, feedback to optimization setup, crystal harvesting, mounting, and cryocooling. As mentioned above, one of the less appreciated facts is the rapid accumulation of an enormous amount of experimental results, and a well-curated database is the foundation of any data-mining or machine-learning efforts aiming to provide predictive inferences for crystallization of new proteins.

2.3.1. Basics of Protein Crystallization

From a phenomenological viewpoint, crystallization is phase separation in a thermodynamically metastable supersaturated system under the control of kinetic parameters, with the favorable outcome being the formation of a crystal. Fulfillment of thermodynamic criteria only implies that crystallization is *possible*, i.e., it is a *necessary but not sufficient* condition for crystallization. Whether the thermodynamically possible outcome is realized, depends on the kinetic parameters controlling the process. While the thermodynamic parameters (classifiable into extensive ones like protein or reagent concentrations and intensive ones such as temperature or pH) are easily controlled by the experimenter, we have only limited influence on (and knowledge about) the kinetic parameters such as equilibration rates, molecular association, pre-assembly, nucleation,[52] and growth kinetics.[53] The kinetic parameters are pathway dependent and do change, for example, with crystallization method (a not irrelevant point for the design of a high-throughout crystallization pipeline), and vary even with less obvious factors such as drop size.[54]

From a microscopic perspective, crystals are periodic assemblies of macromolecules with few and relatively weak contacts between mole-

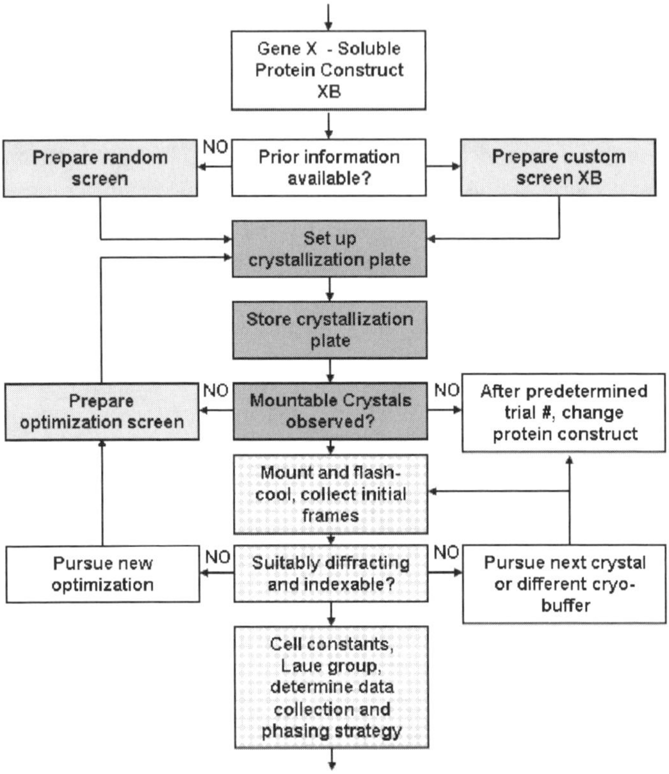

Figure 2.3. Outline of basic tasks and their relation in a crystallization project. Shading indicates basic liquid handling (light gray), crystallization plate setup and handling (dark gray), and mounting and data collation tasks (gray pattern).

cules.[55] Detailed packing analysis of crystal structures revealed that protein–protein crystal contact formation appears to be an essentially stochastic process.[56] Moreover, because current protein structure prediction is not accurate enough nor can protein-solvent interactions be modeled with the necessary precision to pinpoint all contributions to the free energy of crystallization (not to speak of the aforementioned kinetic implications), *ab-initio* crystallization prediction for proteins is not feasible. The absence of *ab-initio* calculations from sequence and physical principles does require statistical, knowledge-based approaches to quantify the likelihood of a certain protein to crystallize under given conditions.

2.3.2. Experimental Crystallization Setup Techniques

In contrast to what one would expect considering the fundamental molecular complexity of the crystallization process as outlined above, the actual setup of the common crystallization experiment is deceptively simple and easy to automate: A weakly buffered protein solution is combined in ratios of order 1:1 with a crystallization cocktail, placed in a more or less closed system and left alone while approaching equilibrium to the point where kinetics permit.

Numerous crystallization techniques have been developed, each with different merits and drawbacks depending on the specific purpose (initial screening, optimization, harvesting, and ease of automation) and type of research environment. **Fig. 2.4** provides a quick overview of a few popular methods and some of their (dis)advantages. Additional special techniques are described in.[31]

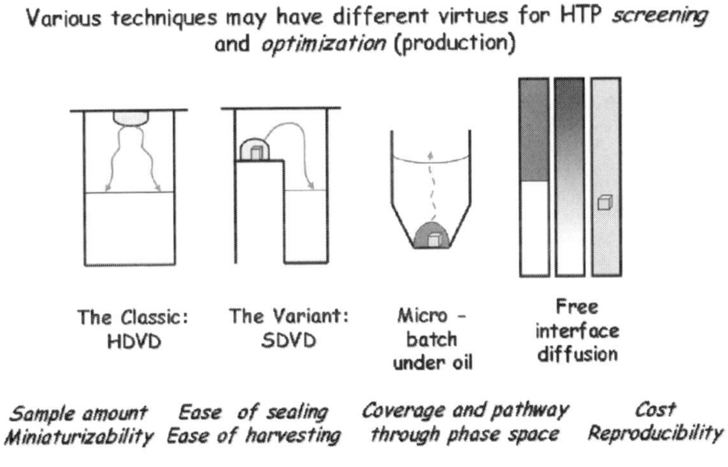

Figure 2.4. Schematics of popular crystallization techniques. Hanging drop vapor diffusion (HDVD) is a most popular method in manual setups; and sitting drop vapor diffusion (SDVD) is commonly used with robotic nanodrop setups. Absence of additional sealing requirements and ease of miniaturization favor automated microbatch screening under oil, although harvesting tends to be more difficult. Use of silicone oils in microbatch also allows partial exchange of solvent vapor, indicated by the broken arrow. Miniaturized free interface diffusion screening chips are gaining popularity, but automation and harvesting issues remain to be resolved. Each method traverses the crystallization phase diagram in a different path and the *same chemical screening conditions* do not necessarily produce the same results.

Maximizing comprehensive sampling of crystallization space with the least amount of material plus ease of miniaturization favor robotic high-throughput setup of nanodrops.[57,58] In addition, absence of sealing requirements makes attractive both microbatch methods under oil[59-62] and free interface diffusion microcrystallization in multilayer soft lithography chips.[63,64] Excluding proteins, which are unlikely to crystallize from expensive expression and purification scale-up, could provide a significant overall efficiency gain—despite possible loss of some proteins due to nontransferability to a different optimization and harvesting technique (**Fig. 2.5**).

Optimization is generally pursued after the major parameters (or factors, in the lingo of experimental design) have been determined during screening. Consistent with the PSI statistics (**Fig. 2.1**.) and literature data, we estimate that only about 10–20% of prokaryotic proteins diffract well out of initial screens, and optimization plays a determining

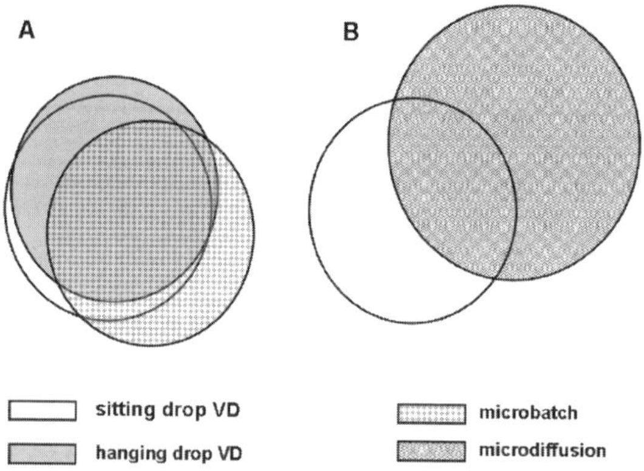

Figure 2.5. Venn diagrams representing crystallization scenarios. (**a**) Overlap in success space between three different techniques. Circles of same diameter indicate equal overall success rate for each method. Overlap between hanging and sitting drop vapor diffusion (VD) techniques is presumably large whereas microbatch may have fewer conditions in common with either.[65] (**b**) Hypothetical scenario representing free interface diffusion microtechnique with potentially higher success rate (larger circle) but limited overlap with sitting drop VD technique. Similar diagrams can be used to visualize the overlap (or lack thereof) of basis sets used in different setups and designs. Regardless of crystallization design used, non-overlap in crucial parameters—either reagents or technical parameters—does make rather difficult unbiased comparison of results between different research groups (or designs). Adapted from the source.[66]

role in overall output (a measure for efficiency) compared to just ramping up throughput.[67,68] From a process-efficiency viewpoint, the use of the same equipment throughout the process can favor deployment of a single technique for screening and optimization,[12] enabling consistent robotic setup optimization experiments subsequent to screening. Some, not yet confirmed, evidence points towards increased screening success rates in TOPAZ (Fluidigm, Inc., www.fluidigm.com) microfluidics chips.[69] From a chemical standpoint, the distinction between crystallization screening and optimization of crystal growth is arbitrary, but a change in method related parameters can significantly impact overlap and success rates (**Fig. 2.5**).

2.3.3. Experimental Design Considerations

Regardless what setup technique is used for crystallization screening, an experimental design should be selected that provides a maximum of information with minimal sample use. Design of Experiments (DoE) for crystallization has been discussed and reviewed.[36] Popular screening protocols such as grid screens,[70] footprint screening,[71] and "sparse matrix" screening[47] have been compared with combinatorial screening in terms of sampling efficiency.[72] Under the assumption of no prior knowledge assisting in the choice of conditions, random (combinatorial) sampling seems most efficient when the successes are sparse and/or clustered. With the accumulation of comprehensive data including protein information as well as negative outcomes, however, one can expect that prior-based predictive models will eventually provide a basis to select key factors for increasing the probability of success beyond random guesses.[36]

Carter[73] provides a concise treatise of optimization techniques such as linear regression, variance analysis, and response surface methods for the analysis of incomplete factorials and related designs (factorial factorials, orthogonal arrays, Hardin-Sloane).[81] Neural networks,[58] and partial least squares, and principal component analysis[74] are also applicable to factorial design optimization. Other multivariate designs used for crystallization optimization are central composite and Box-Behnken design[75] as well as iterative simplex procedures.[76] A brief summary of these designs and their use in high-throughput screening is available.[77] All these designs can be implemented in a HTPX pipeline, and corresponding optimization experiments set up automatically based on initial screening results.

2.3.4. Crystallization Screening as a Sampling Problem

Crystallization success analysis has been treated as a sampling problem of an unknown distribution of successes in crystallization (parameter) space, and detailed analysis and reviews are available.[36,72] Systematic exploration of all parameters (such as reagents, temperature, pH, and a method) is feasible only in low dimensions and with low granularity. If specific parameters selected by intuition or prior knowledge are however dominating factors for crystallization, such designs can rapidly yield valuable information. In their initial 6 × 4 format, these designs are known as grid screen experiments[70] and commercial kits in 96-well format are available. Given that there are about 400 reagents listed in the BMCD,[31,48] a method is needed to effectively distinguish significant factors form nuisance factors that do not contribute to successful outcomes. The key for increasing efficiency *a priori* in all experimental designs is to select all *probable* factors but to *eliminate* irrelevant nuisance factors and *improbable* factor levels (excessive precipitant concentrations, denaturing pH, high PEG concentrations plus high ionic strength,[31] etc.) to reduce dimensionality and volume of the multidimensional parameter space to be sampled.

2.3.5. Robotics and Flexible Design Strategies

One of the activation barriers against advanced screening and optimiza-cific cocktails for each run and/or each protein. Optimal implementation of incomplete factorials, recognized early by Carter and Carter[78] as a systematic means to increase efficiency, requires specific cocktails and buffers to be prepared for each experiment of a run.[93] Unfortunately, when these systematic statistical designs were first introduced, availability of robotics was not as widespread as it is becoming now (a major *factor* contributing to the universal popularity of prefabricated kits). Equally, in random sampling, coverage of the crystallization space is achieved by using each crystallization condition only once, and robotics are a necessity to practically implement the CRYSTOOL random sampling protocol.[79,80] Except for the selection of the basis set, which also contains various detergents,[81] no assumptions about success rate distributions or about factors specific for a particular protein are made. The omission of prior knowledge or absence of assumptions may seem as a serious limitation of random sampling, but there are good reasons not to deviate *prematurely* from the assumption of ignorance. As in the selection of conditions for sparse matrix sampling, the inclusion of prior knowledge—either consciously during analysis, or, quite insidiously, inherent in the experimental design—may affect the estimate of the posterior, in our case predicted, crystallization success probabilities. The formal basis for this reasoning is given by Bayes' theorem[82] and has been discussed elsewhere.[36]

While relative virtues of chemicals comprising a basis set can be readily analyzed, systematic investigation of nonchemical parameters such as drop size, protein concentration, drop ratios, and even temperature, has received less attention,[73] although these method-related variables can be just as influential and valuable parameters as the reagents used.[83] Rapid and miniaturized robotic setup opens opportunities for systematic investigations examining the role of such unexplored parameters.

Given the advantages of custom screen design outlined above, the use of prefabricated kits in high-throughput efforts may appear counterintuitive. Prefabricated screening kits based on previous success analysis have been quite popular,[47] and abundant variations of this first kit are now available (although, as noted before,[84] the rationale for their design is sometimes less than "crystal clear"). Repeated use of such premixed "sparse-matrix" solutions amounts to oversampling of certain spots in the multidimensional crystallization space. Nonetheless, these screens are available in 96-well format suitable for automation and are successfully used even in high-throughput efforts[23] where they allow straightforward basic success rate and propensity analysis.[17,36,85] The conclusions from high-throughput efforts[17,23,85,86] confirm the long-known empirical evidence[87] that PEGs as a class including PEG MMEs[88] exhibit a significantly higher crystallization propensity around physiological pH. HTPX efforts that use commercially available kits also confirm independently that a large number of screening experiments (thousands of conditions) do not seem to be necessary to crystallize the large majority of proteins (at least for the biased set of highly soluble prokaryotic proteins used). The latter observation agrees with the estimate for the number of trials derived by efficiency analysis beyond which return on investment of time, supplies and valuable protein during further screening drastically diminishes. For the average soluble protein based on available frequency and success rate data, it has been estimated that 288 (3×96) trials should suffice to find crystallization conditions with high probability.[72] Beyond this point, the option of protein engineering[24-26] or search for orthologs should be investigated as viable alternative to wasting resources on continued HTPX screening.

2.4. Implementation of Robotic Technology—General Considerations

Operations in high-throughput production mode require a distinctly different skill (and mind) set than classic, academic research usually is experienced at. The wide range of disciplines involved in structural genomics necessitates collaboration in diverse fields ranging from bioinformatics, protein engineering, and robotics development to information technology, statistics, and computer science. In particular, robotic engineering as well as data processing and tracking can grow into formidable tasks not easily handled in a traditional academic bioscience department, compounded with the scheduling and delivery demands of a high-throughput production environment. In publicly funded efforts it is difficult to argue that larger up-front investments such as advanced robotics or information technology infrastructure improvements will provide long-term cost savings over the duration of the project and will allow more efficient use of the available human resources (graduate student talent can be engaged in intellectually more satisfying activity than high-throughput pipetting of crystallization cocktails). **Table 2.1** summarizes a few of the distinctions between academic and industrial environment, each necessitating a different approach to automation.

2.4.1. Efficiency versus Throughput

Irrespective of the environmental and resource boundary of the research environment conditions discussed later in detail, a simple but instructive model can demonstrate in which ways automation can affect the efficiency of a high-throughput screening laboratory. We define overall efficiency E of our process in a simple model as

$$E = \frac{T \cdot S}{C},$$

where T stands for throughput, S for success rate, and C for cost.[12] Given this simple relation, automation, for example, could be deployed with the major objective of a brute force increase in throughput, but likely at great cost. Given cost as a modest and limited constraint in a noncommercial environment, T and S are viable candidates to increase E, the ultimate measure one could choose as an academic shareholder value equivalent. It has been pointed out that in automated crystallization throughput and output do not necessarily equate,[68,89] and S needs to be increased. Robotics can be an efficient means of increasing S by providing optimized, prior-based custom cocktails for better screening designs or aggressive optimization. Particularly in commercial environments, robotics can also be deployed in an effort to reduce personnel cost, an old favorite of corporate management. In contrast, throughput can also scale linearly with cost, as in the low-risk approach taken frequently in academics: adding more graduate students might achieve a proportionally higher throughput, but does not increase efficiency. Although our simple linear efficiency model cannot claim accuracy (in general T and S are specific functions of C, and efficiency models tends to become nonlinear),[90] it is a good reminder to keep the most important factors affecting efficiency and their interactions in mind when planning the deployment of robotics.

Automation can also greatly increase process efficiency as an enabling technology for comprehensive screening and miniaturization. Large enough sample sets can then be screened with little material, and statistically sound go/no-go decisions can be made already early in the process. Failures in later stages tend to become progressively more expensive (**Fig. 2.6**).

The pursuit of a target already showing warning signs of limited success may be heroic but not efficient, depending on the overall goal of

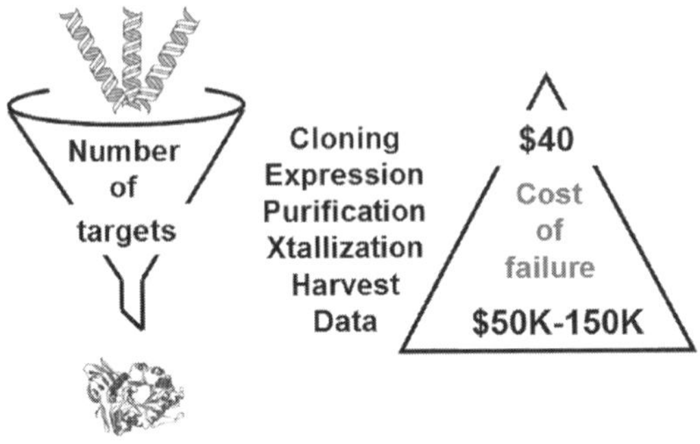

Figure 2.6. Target attrition versus cost of failure. The later a target is abandoned in the process, the more costly steps it has passed through and the more resources it has consumed. As extreme examples, a failure to clone a target out of DNA in high-throughput mode is probably limited to waste of a primer pair and some chemicals and a small amount of labor. The other extreme would be a data set that cannot be phased. At this point, the target has passed all cost intensive experimental stages and accumulated maximal value. Similarly, loosing crystals during cryosoaking is very expensive, affirming that more effort should be spent on systematic investigation of cryoprotection.[91]

the project. Early go/no-go decisions are common in commercial ventures, and academic examples are the "two-tiered" approaches to crystallization[86,92] or the estimate of a point of diminishing return in random screening based on statistical analysis.[72] A classical example from pharmaceutical industry emphasizing the importance of early detection or prediction of potential problems is the frequent failure of new therapeutic drugs in late phases of clinical trials due to unexpected drug interactions.[93]

2.4.2. Patent and Licensing Issues

An issue that affects commercial ventures usually more than academics is the need for licensing of patented materials, processes, or copyrighted code. In the USA, for example, the use of nanoliter-drop technology in crystallization is protected and was the subject of an infringement dispute involving the patent holder Syrrx in San Diego, Calif., (now owned by Takeda, http://www.takedasd.com/) and Oculus Pharmaceuticals, Birmingham, Ala. (U.S. Patent 6,296,673 "Methods and apparatus for performing array microcrystallizations"). Similar issues may arise from the U.S. patent 6,860,940 granted to the University of California regarding the use of "Automated macromolecular crystallization screening." Academic laboratories are usually exempt or at least not pursued when infringing on the patents (due to low probability of any financial recovery), but given the ever increasing intertwinement of academics and commercial contract work, patent and licensing issues cannot be ignored.

2.4.3. Service and Maintenance

Vendor or third-party on-site service contracts constitute a nearly mandatory but easily forgotten expense for complex equipment. Particularly, if no dedicated internal service or engineering team is available, service contracts (usually due after the warranty period) can provide a good return on investment. On-site service contracts in general seem to cost about 10% of the equipment price per year. Their value for any production schedule is evident; in very small laboratories and with basic equipment it may be possible to rely on local talent.

2.5. Implementation of Robotics in the Academic Laboratory

Let us assume we have the task of automating a protein crystallography laboratory. Picture an academic environment with limited resources, likely from different funding sources with an equipment budget for crystallization automation somewhere in the range of $100K to $500K. As academic funding is highly fluctuating, a primary consideration might be the necessity to acquire systems that are modular and can be extended once new funding becomes available. Based on such considerations, one of the goals of the TB Structural Genomics Consortium crystallization facility, funded as a part of a NIH-sponsored PSI-I centers, was the development of affordable and modular high-throughput robotics.[12,79] Compatibility of components becomes an important issue in modular design. While SBS (Society for Biomolecular Screening) labware standards have been adopted throughout the automation community, a major obstacle to combining products from multiple vendors is poor interoperability and compatibility of the control software. Proprietary controls and drivers, undocumented interfaces, and nonstandard data formats and databases create major nuisances. In addition, for reasons not entirely clear, companies tend to overestimate the cash value of their respective software, and academics are usually quite capable of writing their own interfaces if needed. Irrespective of the specifics discussed later, a primary question whenever a modular and extensible design is desired, are hardware and software flexibility and adaptability. In an academic multiuser environment, qualities like robustness of the hardware and simplicity (and fail-safe mode) of controls might also be considered.

2.5.1. Objectives and Cost

In academics, a primary objective of automation might be to relieve graduate student and postdoc talent from intellectually modestly challenging activities like high-speed pipetting, cocktail mixing, and plate shuffling to accommodate more productive activities (such as repairing robots, etc.). Raw throughput is generally a minor concern; even modest robotic crystallization equipment probably outstrips the capacity of the up-front protein production (most proteins that are difficult to crystallize are also those that are hard to produce—ordered phase separation, after all, requires solubility to begin with).

Miniaturization, reproducibility, and convenience are probably the next most significant benefits of robotic automation in the small laboratory, in addition to easier and consistent record keeping via automated capturing of the experimental data (the need for effective data management has been discussed already). At any rate, the deployment costs for robotics should make allowance for database integration, and if not yet implemented, this would be an excellent point in time to invest in a LIMS system for the whole laboratory. The initial cost and activation barrier for automation may be greater than for the simple "grad student" model and a long-term view of the benefits and of investment amortization may be necessary (**Fig. 2.7**). The step into robotics could generally provide a unique opportunity for a critical process review across the whole laboratory.

2.5.2. Bottlenecks

A crucial part in the evaluation of throughput requirements or expectations is to define the rate-limiting steps. In a crystallization process flow **(Figs. 2.2 and 2.3),** the total throughput is determined by rate-limiting

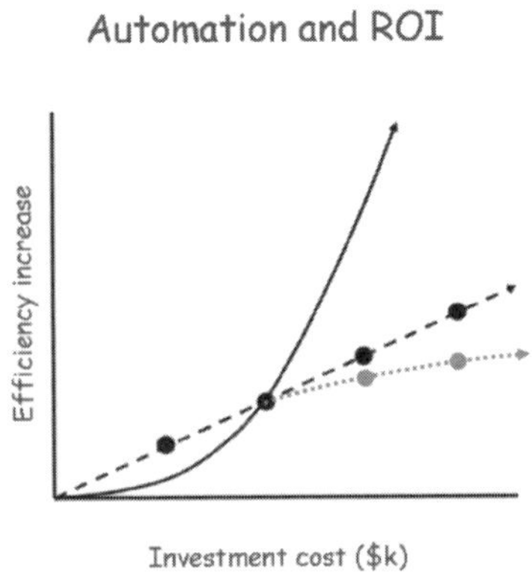

Figure 2.7. Comparison of investment models. The academic graduate-student model (*n* students = *n* times throughput, dashed line) scales linearly, while robotic deployment (full line) requires greater up-front investment but later outperforms the graduate students. Note that although it is possible to hire a student 50% time, half a robot is seldom a purchase option. Note also that a realistic graduate-student model (dotted line) does not scale linearly due to collisions of the students in the overcrowded lab—a typical example for creating a (verbatim) new upstream bottleneck.

steps or bottlenecks. Due to deployment of robotics and the development of new methods and techniques, rate-limiting steps do shift and some extrapolation regarding the next expected rate limitations is advisable while addressing the current one. Without paralleling measures, increased robotic protein crystallization throughput will likely outstrip the protein production capacity,[95] providing an example for *upstream* shift of a rate-limiting process step. Observation of crystallization experiments exemplifies a new bottleneck created *downstream*: given sufficient protein supply, it is quite conceivable—even with affordable equipment—to set up about twenty to thirty 96-well crystallization trays per eight-hour shift. At 20 plates/day capacity and with a plate shelf life time of four months and eight observations in that period, roughly 2400 plates accumulate and require at least one observation every two seconds. Given the rapid kinetics and equilibration in nanodrops,[54,57] a six-day shelf life and daily observation, for example, accumulate less plates (120) but still require one observation every two seconds. Cleary, image acquisition and analysis of outcomes must also be highly automated, and corresponding action in response to the outcome, such as scheduling of mounting or automated setup of optimization and harvesting plates, needs to follow in a timely manner (crystals do not last indefinitely and may degrade or disappear with time). In addition, with no automated plate-handling system, at least one person would be occupied shuffling (and tracking) plates between the various stages of the setup.

The above scenario serves to impress that high crystallization throughput of even a few proteins per day is not sustainable without careful planning of the entire automation setup and thorough operations review. A singular deployment of high-speed robotics, in particular, if intended to significantly increase throughput, will likely create new bottlenecks downstream. If efficiency considerations like relief of personnel from repetitive tasks or reproducibility and standardization of the setup are key goals, capacity related downstream backups may be less of a concern.

2.5.3. Kits versus Custom Cocktails

A much discussed issue is the question whether to use fixed kits of the sparse-matrix type or its numerous descendants (in an attempt to avoid robotics for cocktail preparation) or to prepare and use custom cocktails. The cost of ready-to-use kits in 96-well format using the above 20 plates/day scenario amounts to about $80K/year (100 μL reservoirs) and could be about cut in half by manual re-arraying of conventional 10-mL kits. For low throughput, re-arraying might be feasible, but several arguments already discussed in the "Experimental Design" section speak against a setup without capability to produce custom cocktails. First, despite screening with kits has been proven successful,[17,23] optimization will require preparation of specific cocktails. In addition, prior information about protein properties often precludes certain conditions (extreme pH, high salts, certain metals, etc.) while statistical analysis may indicate that a particular class of proteins may prefer certain crystallization conditions. Methods like true random screening[80] invariably imply setup of new crystallization cocktails, a requirement that probably has hampered the widespread deployment of factorial designs as well.[73,96]

2.5.4. Limitations to Full Automation

Full walkaway automation up to but not including harvesting is conceivable giving the current equipment on the market and has been demonstrated (albeit at substantial cost) in several custom-made industrial designs. Cocktail preparation, plate setup, automated crystal recognition, and subsequent optimization can be integrated with plate-handling robotics and provide no principal (nonetheless financial) challenges. Process automation currently stops at the harvesting stage, largely due to the expense of micromanipulation, and the need for rather advanced machine-vision tools to allow real-time processing of the events during mounting. However, new reproducibly manufactured mounting loop designs[97,98] and micromanipulation actuators for robots (Square One Systems Design, http://www.sqr-1.com/, in corporation with Bruker AXS, http://www.bruker-axs.de/) are being developed, and collaborative efforts are underway to address this remaining manual bottleneck. For an academic effort, the chance that crystal harvesting becomes a true bottleneck is low and, in fact, would vindicate the preceding automation efforts and provide a strong indication for success. For commercial HTP crystallographic screening of drug target–ligand complexes, for example, this scenario may be quite different,[6] and effective crystal harvesting has become a bottleneck.

2.5.5. Basic Automation Scenarios

Based on the assumption of modest throughput requirements and no necessity for full walkaway automation, two low-budget approaches to automation are conceivable: (1) selection of a single system that can prepare both cocktails (perhaps in a limited fashion) as well as set up the crystallization plates or (2) a basic dual station layout using separate cocktail preparation with a generic liquid handling system followed by a dedicated plate-setup robot. The major reason for separating plate setup from cocktail production is different requirements for dispensing precision, volume, and speed. Fast, small (μL to nL) volume and very crystallization accurate (also in geometric terms) dispensing is mandatory for plate setup, whereas large volume (mL) handling, with modest speed and precision requirements, suffices for cocktail production. However, with a limited number of tips and a large chemical basis set, the "traveling salesman problem" of the tips may need some optimization to prevent cocktail preparation from becoming a rate-limiting step. Proper arrangement and dispensing sequence of reagents can be first steps to increase dispensing speed.

Another advantage of the separation between the cocktail stage and the plate setup is that simple one-to-one dispensing into reservoir wells and into drop aliquots suffices, once the cocktails are produced in a 96-well format. In addition, compared to a single-stage setup, failure of one component does not affect the other. For example, advance cocktail production can continue while the plate-setup robotics are inoperative.

2.6. Instrumentation

2.6.1. Liquid Handling Stations

Numerous vendors of multipurpose, generic liquid handling stations often integrate their systems with plate handling and occasionally even with image acquisition (discussed under Section "Fully Integrated Systems" below). The basic requirements for a liquid handler include variable span tips, liquid sensing capability, controllable aspiration speed, and software that processes a custom work list of instructions. Disposable, coated tips capable of liquid sensing are usually too expensive in high-throughput mode, while wash cycles of fixed tips can be a major contribution to the total preparation time. More tips are better, and eight tips seem to be a standard configuration (if disposable tips are desired for any reason, the Mosquito robot (Molecular Dimensions Ltd., Soham, England) may provide an alternative). While both the volume and mechanical precision requirements for cocktail preparation are relatively easy to fulfill, the need for high positional precision and accuracy becomes important when attempting to set up crystallization plates with the same piece of equipment. The positional precision to place nL-size drops reliably onto each other needs to be better than about ±100 μm. Liquid handling robots suitable for cocktail preparation are available from Packard Instruments/Perkin Elmer, http://las.perkinelmer.com/), Tecan (**Fig. 2.8**), http://www.tecan.com/, Biomek® (Beckmann Coulter, http://www.beckmancoulter.com/), Zymark (http://www.caliperls.com/), Hamilton (http://www.hamiltoncomp.com/), Gilson (former Cyberlab; Middleton, Wisc., http://www.gilson.com/), Cartesian (http://www.cartesiantech.com/protein_crystallization.htm), Zinsser Analytical (http://www.zinsser-analytic.com/) and many others. Depending on the options (a bar-code reader being an inexpensive but important one), which often include basic labware storage and plate loading, such liquid handlers cost about $100–200K. As already mentioned, quality and compatibility of the software as well as database and LIMS interfacing may be more of a decision criterion than the mechanical quality, which is generally *en par* between vendors (the competition in the generic liquid handling market is stiff, and negation skills can be quite important).

Figure 2.8. Tecan Freedom EVO® 100 liquid handling robot. Its small footprint and modest price make it a suitable platform for generating crystal screens to complement a robotic plate setup system. Systems like this also allow the flexibility to introduce ligand screening into the crystallization pipeline. Image courtesy of Center for Structural Biology at Texas A&M University.

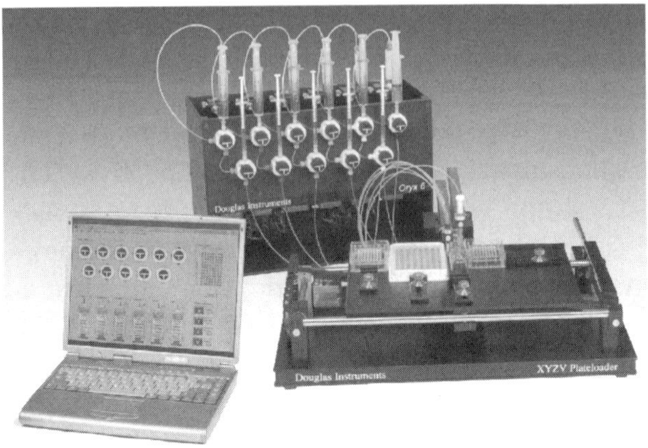

Figure 2.9. The Douglas Instruments' ORYX 6. The Douglas IMPAX and ORYX robots were designed primarily for microbatch crystallization where small drops (down to 200 nL) are dispensed under oil, but the ORYX can also set up sitting drop experiments in 96-well plates. The target crystallization plates are at the extreme left and right. Stock solutions are in the 96-well plate with a white frame near the center of the XY stage. Basic software for both screening and optimization—including efficient multivariate designs for up to four dimensions—is included. Image courtesy of Douglas Instruments Ltd., http://www.douglas.co.uk/.

2.6.3. Liquid Handling Stations

Microbatch. The first in fact usable and affordable commercial single stage robots were the IMPAX and than ORYX series of crystallization robots (Douglas Instruments, Berkshire, United Kingdom, http://www.douglas.co.uk/, **Fig. 2.9**), which are eminently suitable for microbatch crystallization and which can also prepare sitting or hanging drop experiments using special labware. The benefits of microbatch have been extensively reviewed, [62,65] and an obvious advantage is the absence of additional plate sealing requirements. Harvesting, on the other hand, can be tricky, and change in technique from microbatch to vapor diffusion can change pathway-dependent kinetics and phase-space coverage (**Fig. 2.5**). The Douglas robots also provide software support for several optimization schemes, including the option for custom designs.[75] For limited budget (around $60K) and modest throughput, the Douglas robots and microbatch techniques are a viable automation option. The microbatch technique has also been used extensively for rapid high-throughput crystallization screening using a Hydra multisyringe dispensing robot integrated with an automated image acquisition stage.[60,99]

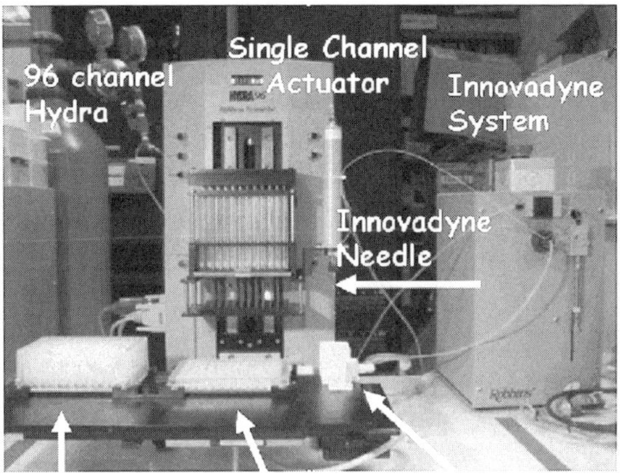

Figure 2.10. The Hydra Plus-One prototype as developed in collaboration of the macromolecular crystallography group at LLNL and Apogent Technologies.[100] Information about the latest development built by Art Robbins Instruments is available from http://artrobbins.us/. Image courtesy Texas A&M University.

Sitting drops. A simple but reliable solution to crystallization plate setup in sitting drops is exemplified by the Hydra Plus-One system (**Fig. 2.10**) consisting of the proven Hydra multichannel syringe dispenser augmented with a contactless, single channel Innovadyne™ microsolenoid dispenser.[100] A later version of the system is sold by Matrix Technologies (http://www.matrixtechcorp.com/) as the Hydra® II Plus-One System, and a cost-effective but improved system (Phoenix) is available from Art Robbins Instruments (http://www.artrobbins.us/). The Hydra-based systems dispense cocktail reservoirs and cocktail drops rapidly by 1:1 syringe transfer from premixed solutions in 96-well blocks. The protein is dispensed without re-arraying losses from a single or multichannel microsolenoid dispenser into the precipitant aliquots. The whole process of plate setup can be accomplished in 60–90 seconds, with sufficient time for wash steps after the plates have been sealed. The manufacturers claim usability down to 100 nL (on glass slides), but a problem with the syringe technique is that touching-off of the drops at such small volumes into the crystallization plates is hardly achievable in routine use due to limited precision of both plates and needle height adjustment. Drop sizes of 500–200 nL seem to be the practical limits. Steel needles are claimed to provide higher precision, but in a multi-user environment, the robust DuraFlex needles are definitely an advantage. Despite all engineering controls, the possibilities of misaligning a crystallization plate seem to be endless.

A distinct limitation of multisyringe-based robots is that all 96 channels dispense the same volume, which precludes the possibility of drop ratio variations across the 96-spot array. Independent variation of dispensing parameter including drops is a distinct advantage of more flexible systems such as the Innovadyne 96+8 screen maker.

The Innovadyne ScreenMaker™ 96+8 (http://www.innovadyne.com/) extends the simple concept of the Hydra Plus-One to a flexible and versatile design (**Fig. 2.11**). The ScreenMaker has a standard array of 96 independent contactless microsolenoid dispense needles plus a separate eight-channel array for protein dispensing. The large aspiration volume and dynamic range enable independent high-speed noncontact dispensing in the low-nanoliter to microliter range. Variation of the drop-size ratios over the whole array (**Fig. 2.11**) is a valuable but commonly not explored feature in optimization. A large aspiration volume and dynamic range also allows preparing custom cocktails on demand, again a feature not available with the fixed syringe systems. At present viscosities corresponding to 30% PEG 8000 solutions can be handled, assuring substantial coverage of the crystallization space. The ScreenMaker can pick up 96 stock reagents and perform cocktail setup (with average 2.5 transfers/well) in about 15 minutes and afterwards dispense the aliquots and protein in about 45 seconds using a single needle and in 15 seconds if the protein is re-arrayed into all eight separate heads. Drops down to 100–50 nL can be set up using this instrument. The ScreenMaker 96+8 has

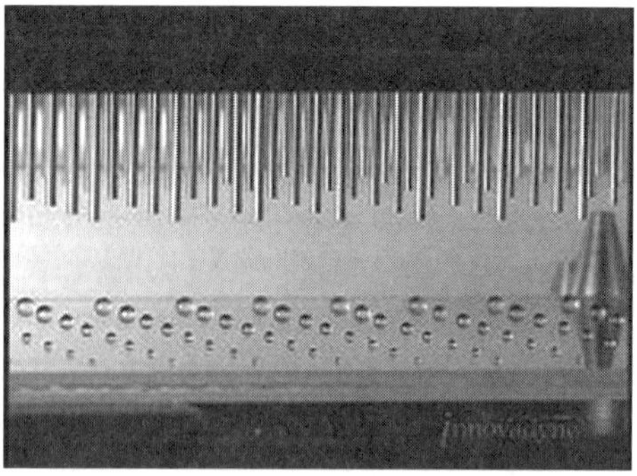

Figure 2.11. Droplet pattern created by independently controllable, contactless 96-channel Innovadyne dispensing system. Such systems are suitable for nanoliter plate setup as well as cocktail preparation in the 100 μL range. Image courtesy of Innovadyne Technologies (http://www.innovadyne.com/).

a street price of slightly below $200K, comparable to a combination of a basic liquid handling robot ($100K and higher) and a Hydra-II Plus-One ($80K).

Similar contactless high-speed dispensing technology is offered by Cartesian Systems, including custom designed complete solutions for cocktail preparation and protein dispensing in the upper price range.

2.6.3. Sealing

Automated plate sealing must be integrated in the process for full walk-away automation. In the absence of integrated plate handling the robot operator, who also needs to supply the machines with consumables and labware, can seal plates manually in "semi-automated" mode. Cocktails are commonly prepared (or sold) in 96-well deep blocks containing 1–2 mL of reagent and can be sealed with thermal adhesive tape using a cheap manual heat sealer. Short-time delays between preparation and sealing are also uncritical. Crystallization plates in contrast need to be sealed rapidly to avoid evaporation losses with small drops, and the heat transfer to the drops precludes thermal sealing. Pressure-activated adhesive tapes and precut sheets are available, and most automated sealers can adequately handle tape on large rolls. Sealers in the price range of $20–30K are available from Brandel (http://www.brandel.com/) and several other lab-automation companies (see www.corning.com for a summary). Note again that the absence of sealing requirements in microbatch techniques provides a viable option for smaller laboratories, or for cases when rapid initial screening is the major objective.[60]

2.6.4. Visualization

Crystallization laboratories quantify their crystallization results usually by assigning a quality score on a scale with a granularity of about 10. With proper training and crossvalidation, individuals as well as automated image recognition and scoring routines[13,37,79,99,101-103] can assign scores in a relatively consistent way *within* a given laboratory. Unfortunately, what the scores really mean varies, and no strong common scoring metric exists. The negative consequences for inter-laboratory comparison of otherwise potentially interesting statistics such as relative success rates for proteins from different organism[17] have been pointed out.[36]

A number of imaging stations are offered, ranging from rather basic systems in the $40–60K range (Diversified Scientific's CrystalScore™; TriTek/Molecular Dimensions' CrystalPro™; RoboDesign/Rigaku's Minstrel-I; deCODE Genetics' Crystal Monitor) to integrated imaging and storage solutions that cost $250K or more (Veeco's OASIS 1750 system; RoboDesign/Rigaku's CrystalMation™; Bruker/Discovery Partners' CrystalFarm™ (**Fig. 2.12**)). While at least in the more expensive systems, the mechanical parts as well as image acquisition and quality are decent, the image analysis software often tends to be the weak point. Image analysis software needs to do more than just display the drops in order to have an effect on the process integration. Both false negatives and false positives are problematic: Flagging indiscriminately every feature in a drop in order not to loose any potential crystals is counterproductive. Most commercial software tends to err on the conservative side, and particularly features located at drop perimeters are commonly detected false positives, necessitating human inspection (or automatic pursuit) of too many irrelevant observations. In a high-throughput environment the objective is not to pursue every possible target to the bitter end but to efficiently advance the ones with highest probability of success. The loss of low-probability targets is largely compensated by more comprehensive screening possible through automation. The fear of loosing the only crystal (or "promising conditions") ever observed seems to besiege the academic community in particular, and may be the driver for commercial software developers to err on the conservative side. On the other hand, of course, too many false negatives (missing crystals) also are unacceptable. Good software will be trainable so that the most efficient compromise suitable for the given expectations can be achieved. Crystal image analysis software has been developed by several academic groups,[37,101,102] and the software developed at the TB Structural Genomics Consortium as recently described[13] has been licensed to Axygen (http://www.axygen.com/).

2.6.5. Plate Handling and Storage

To combine separately acquired liquid handling, plate setup, sealing, and imaging modules and also to combine them with storage, automated plate handling and tracking are required (that fact that plate tracking through the process requires adequate IT and database structure has already been emphasized). In practically all cases, the system will include bar code tracking to allow the system scheduler and database to track the position and flow of the plates through the process.

Probably the first step where plate handling effectively reduces workload is the loading of automated imaging stations. If plates are stacked in racks, a basic loading robot with a rotating arm such as the PlateCrane models (Hudson Controls, http://www.hudsonwww. hudsoncontrol.com/, Springfield, N.J.) can be used to place plates on the deck of the imaging station. At a modest acquisition rate of 6 s/well, one plate can be viewed in approximately 10 minutes, and in 8 hours about 48 plates need to be placed and removed. Depending on the plates used, this corresponds to one or two storage racks for the PlateCrane. A drawback of the simple systems is the limited working envelope, at a price of about $30K.

Figure 2.12. The Bruker/Disovery Partners CrystalFarm™ exemplifying the integrated storage incubator and imaging systems available (http://www.bruker-axs.de/). The camera, visible in the cutout, is enclosed in the vibration insulated incubator with a capacity of 400 standard SBS footprint plates.

A general concern is also the temperature stability of the plates during storage and observations. Although there still are no convincing arguments in favor of a certain temperature as far as overall crystallization success is concerned, large changes between observation and storage conditions are not desirable. In vapor-diffusion plates, the volatile components tend to condense on the sealing tape when a plate is stored at a lower temperature than the observation takes place. For that reason, at least the imaging station and the plate storage tend to be contained in a closed environment, and the more expensive integrated imaging stations (**Fig. 2.12**) follow that concept using incubators, sometimes at the cost of limited compatibility with other systems and platforms. Basic incubators are available from VWR Scientific (http://www.vwrsp.com/), vibration isolated models from Molecular Dimen-sions (http://www.moleculardimensions.com/), RuMed (http://www.rumed.de/), and Kendro/Thermo (http://www.kendro.com/).

A system enclosure housing a robot with a large envelope is a viable alternative, and industrial type, multiaxis robots on linear track provide a robust and surprisingly affordable solution. Plate enclosures from several hundred to thousands of plates have been built, largely in biotech or structural genomics ventures as well as in synchrotron based user facilities (**Fig. 2.13**).[6,85] In fully automated laboratories, humidity con-

Figure 2.13. The EMBL Outstation Hamburg, a synchrotron based user facility, installed a custom integrated, fully automated visualization and storage solution. The system can store up to 10,000 crystallization plates and can inspect up to four plates simultaneously in 4 minutes. Image courtesy of Jochen Müller-Diekmann, EMBL Outstation Hamburg, Germany (http://www.embl-hamburg.de/services/facilities/crystallisation/index.html).

trol is also occasionally incorporated to slow the evaporation process during nanodrop setup. The plate-setup robot is then also placed into the enclosure, but given the highly varying partial pressures of cocktail components (e.g., isopropanol vs. PEG 10000), this may not be as effective as it appears. Rapid setup (a minute or less) followed by immediate sealing seems to work satisfactory for drop sizes of about 200 nL or above. Compared to the simple, PlateCrane type robots, the much increased envelope, payload, and ruggedness of the industrial multiaxis robots like the MotoMan SV3 series or equivalents from Misubishi, Epson, or Stäubli come at a reasonable price ($60–80K including eight-feet linear track for the SV3). As always, open software interfaces and documentation are important, and considerable effort in software integration and writing the custom process protocols needs to be anticipated.

2.6.6. Fully Integrated Systems

Integrated protein crystallization systems range from customized liquid handling stations from a single manufacturer to fully automated, laboratory-size integrated automation solutions offered by systems integrators (for example, the Rhombix OPUS system, http://www.rhombix.com/). Most of the larger companies selling liquid handling robotics systems are custom-designed to specifications, and specialized integrators/developers can also be hired to design and deploy integrated systems. Members of the Novartis Research Foundation, the Scripps Research Institute, and Syrrx (now Takeda, http://www.takedasd.com/) have codesigned Agincourt™, one of the first truly integrated nanodrop crystallization systems using Cartesian Technologies' SynQuad's system.[5] In similar collaborations, DeLucas (University of Alabama, Birmingham) and his team of industrial partners have developed a highly automated platform[58] used by one of the PSI-I structural genomics centers.

2.6.7. Cryoprotection

Rapid cooling of crystals to cryogenic temperatures (quenching or flash-cooling) has become a standard procedure in macromolecular crystallography.[29] The foremost reason for cryocooling is the drastic reduction of radiation damage,[28] which eliminates the need for multiple crystals and the associated data merging errors; and secondly, increasing the resolution due to reduced thermal vibrations. An additional benefit is that, once the crystals are cryoprotected and safely embedded in the solid amorphous mother liquor of their loops, they can be handled quite easily and reliably by high-throughput mounting robots.

Successful cryocooling depends on a number of factors, only few of them well established under controlled conditions. If the crystallization cocktail does not *a priori* contain high enough concentrations of reagents like PEG, MPD, or glycerol to prevent freezing (i.e., the formation of ice destroying the crystal), the mounted crystal needs to be swiped through a cryoprotectant before being flash cooled. For reasons not entirely clear, even rapid quenching and amorphous state of mother liquor do not guarantee successful cryocooling. Excessive increase in mosaicity and loss of resolution are common mishaps. A number of annealing procedures occasionally reducing mosaicity have been reported,[27,104,105] but have not yet become established standard procedure in HTPX pipelines. Although a common recipe for cryoprotectants is to spike mother liquor with glycerol, PEG, MPD, or other additives, few systematic studies of generally applicable procedures for high-throughput efforts appear in the literature.[106] Clearly, these methods are of great relevance to the objectives of high-throughput crystallization, but they are time consuming and risky for the crystals, and problems during soaking and cryoprotection are probably the single most significant source of loss of crystals in both high- and low-throughput crystallography. The advent of precisely fabricated and stable cryoloop designs,[97,98] universal micromanipulation robots (Square One Systems Design, Jackson, Wyo., and Bruker AXS, Madison, Wis.) combined with advanced machine-vision tools allow automated crystal harvesting (**Fig. 2.14**) and will remove the last manual bottlenecks towards fully automated high-throughput crystallography.[111]

2.6.8. Robotic Sample Mounting

Once the crystals are safely cryoprotected in their mounting loops under liqud nitrogen, the further handling is quite amenable to automation. Robotic sample mounting has become a *de facto* standard on most HTPX synchrotron beam lines, and the Hampton-style magnetic bases are widely used. Practically every major synchrotron facility and larger biotech companies have developed mounting robots for their HTPX

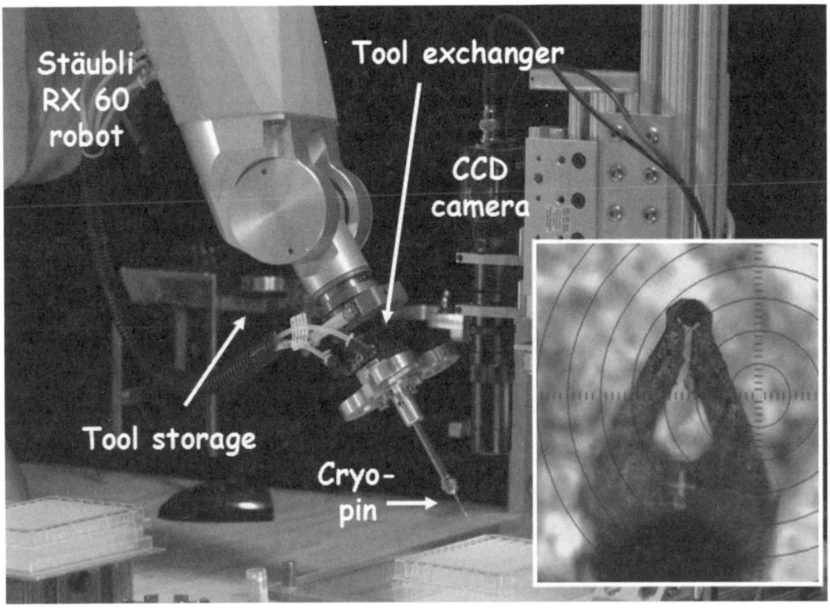

Figure 2.14. Prototype of crystal harvesting robot. An anthropomorphic six-axis robot is equipped with a versatile, pneumatic tool exchange end effector, which can pick up cryoloops, knives, liquid exchange nozzles, and other custom tools. Controls in a game-style keyboard layout and machine vision support enable reproducible and fast machine assisted crystal harvesting and quenching. The insert shows a robot-harvested, ~20 μm large, dyed crystal in a microfabricated MiTeGen cryoloop. Image courtesy q.e.d. life science discoveries, Livermore, CA, and Square One Systems Design, Jackson, Wyo.

beam lines (for example,[79,107-110]). In addition to allowing more efficient use of valuable beam time, robotic mounting also allows the safe removal and re-storage of a sample, should the initial diffraction snapshots indicate doubtful crystal quality. Potentially better crystals can be then mounted and examined without risk of loosing the perhaps best one found so far.

Commercial systems, which are also suitable for in-house lab sources, are available (Mar Research (http://www.marresearch.com/), Rigaku-MSC (http://www.rigakumsc.com/), Bruker-AXS (http://www.bru-ker-axs.de/)). Most commercial mounting robots are packaged together with the top-of-the-line diffractometer, and actual quotes vary depending on negotiation skills. As a ballpark figure, $200–250K need to be additionally budgeted for a sample mounting robot to be included in complete diffractometer system packages.

2.7. Conclusions

Automation in protein crystallization is becoming increasingly attractive, but for a successful deployment in any laboratory settings, the expected benefits must be offset against the total cost of ownership. Although basic equipment is affordable, there is more to automation than just purchasing a liquid handler and/or plate incubator with integrated camera. Automation for high throughput should parallel a change in mindset, including a re-analysis of the whole process of crystallization and protein production. Robotic high-throughput crystallization screening in microdrops allows in an unprecedented way to implement strategies that identify promising leads early, and provides opportunity for corrective action at the protein construct level—before more resources are expended on low-probability targets likely to fail in later steps.

High-throughput crystallography is not just structural biology on steroids, and high-throughput crystallization of even a few proteins per day is not sustainable without careful planning of the entire automation setup and thorough operations review. A singular deployment of high-speed robotics, in particular if intended to significantly increase throughput, will likely create new bottlenecks downstream. Data capture, warehousing, and curating is of enormous importance not just for process control but also for successful data mining of highly dimensional and complex proteomics data. If proteomics and crystallization data analysis is to evolve from basic frequency and propensity analysis to truly predictive models of specific inference, it is mandatory that common metrics for scores and minimal standards for experimental design are maintained. Realistic planning and full consideration of high-throughput process and design principles will go a long way to accomplish a successful and financially sound transition into robotic high-throughput screening.

Acknowledgements

Li Chen (RCSB Rutgers) has helped extracting information from the PSI target database. Parts of the instrumentation development reviewed in this publication was funded by NIH grants P50 GM62410 (TB Structural Genomics) and R41 GM073278.

References

1. Norvell JC and Zapp-Machalek A. **Structural genomics programs at the US National Institute of General Medical Sciences**. *Nat Struct Biol Suppl* 2000, **7**:931.

2. Blundell TL, Jhoti H, and Abell C. **High-throughput crystallography for lead discovery in drug design**. *Nat Rev Drug Disc* 2002, **1**:45–54.

3. Harris T. **The commercial use of structural genomics**. *Drug Disc Today* 2001, **6(22)**:1148.

4. Rees DC, Congreve M, Murray CW, and Carr R. **Fragment-based lead discovery.** *Nature Rev Drug Discov* 2004, **3**:660–672.

5. Weselak M, Patch MG, Selby TL, Knebel G, and Stevens R. **Robotics for automated crystal formation and analysis**. *Meth Enzymol* 2003, **368**:45–76.

6. Burley S. **The FAST and the curios**. *Modern Drug Disc* 2004, **7**:53–56.

7. Knowles J and Gromo G. **Target selection in drug discovery**. *Nature Rev Drug Disc* 2003, **2**:63–69.

8. Kantardjieff K and Rupp B. **Structural bioinformatics apporaches to the discovery of new antimycobacterial drugs.** *Curr Pharmaceutical Design* 2004, **10**:3165–3211.

9. Cavasotto CN, Orry AJ, and Abagyan RA. **Structure-based identification of binding sites, native ligands and potential inhibitors for G-protein coupled receptors.** *Proteins* 2003, **51**:423–433.

10. Abagyan RA and Totrov MM. **High-throughput docking for lead generation**. *Curr Opin Chem Biol* 2001, **5**:375–382.

11. Yon J and Jhoti H. **High-throughput structural genomics and proteomics: Where are we now?** *Targets* 2003, **2**:201–207.

12. Rupp B. **High throughput crystallography at an affordable cost: The TB Structural Genomics Consortium crystallization facility.** *Acc Chem Res* 2003, **36**:173–181.

13. Segelke B, Schafer J, Coleman M, Lekin T, Toppani D, Skowronek K, Kantardjieff K, and Rupp B. **Laboratory scale structural genomics**. *J Struct Func Genomics* 2004, **5**:147–157.

14. Dry S, McCarthy S, and Harris T. **Structural genomics in the biotechnology sector.** *Nature Struct Biol Suppl* 2000, **7**:946–949.

15. Goodwill KE, Tennant MG, and Stevens RC. **High-throughput X-ray crystallography for structure based drug design**. *Drug Disc Today* 2001, **6(15)**:S113–S118.

16. Heinemann U, Illing G, and Oschkinat H. **High-throughput three-dimensional protein structure determination.** *Curr Opin Biotechnol* 2001, **12**:348–354.

17. Kimber MS, Vallee F, Houston S, Necakov A, Skarina T, Evdokimova E, Beasley S, Christendat D, Savchenko A, Arrowsmith CH, Vedadi M, Gerstein M, and Edwards AM. **Data mining crystallization databases: Knowledge-based approaches to optimize protein crystal screens.** *Proteins* 2003, **51**:562–568.

18. Chen L, Oughtred R, Berman HM, and Westbrook J. **TargetDB: A target registration database for structural genomics projects**. *Bioinformatics* 2004, **DOI:10.1093**:bth300.

19. Loll PJ. **Membrane protein structural biology: The high throughput challenge**. *J Struct Biol* 2003, **142**:144–153.

20. Wiener MC, Verkman AS, Stroud RM, and van Hoek AN. **Mesoscopic surfactant organization and membrane protein crystallization.** *Prot Sci* 2000, **9**:1407–1409.

21. Laible PD, Scott HN, Henry L, and Hanson DK. **Towards higher-throughput membrane protein production for structural genomics initiatives.** *J Struct Func Genomics* 2004, **5**:167–172.

22. Mayer MR, Dailey TA, Baucom CM, Supernak JL, Grady MC, Hawk HE, and Dailey HA. **Expression of human proteins at the Southeast Collaboratory for Structural Genomics.** *J Struct Func Genomics* 2004, **5**:159–165.

23. Hui R and Edwards A. **High-throughput protein crystallization**. *J Struct Biol* 2003, **142**:154–161.

24. Edwards AM, Arrowsmith CH, Christendat D, Dharamsi A, Friesen JD, Greenblatt JF, and Vedadi M. **Protein production: Feeding the crystallographers and NMR spectroscopists**. *Nat Struct Biol Suppl* 2000, **7**:970–972.

25. Waldo GS, Standish BM, Berendzen J, and Terwilliger TC. **Rapid protein-folding assay using green fluorescent protein**. *Nature Biotechnol* 1999, **17**:691–695.

26. Dale GE, Oefner C, and D'Arcy A. **The protein as a variable in protein crystallization**. *J Struct Biol* 2003, **142**:88–97.

27. Kriminski S, Caylor CL, Nonato MC, Finkelstein KD, and Thorne RE. **Flash-cooling and annealing of protein crystals**. *Acta Crystallogr* 2002, **D58**:459–471.

28. Garman E and Nave C. **Radiation damage to crystalline biological molecules: Current view**. *J Synchr Rad* 2002, **9**:327–328.

29. Garman E. Cool data: **Quantity AND quality.** *Acta Crystallogr* 1999, **D55**:1641–1653.

30. Patterson SD. **Data analysis—The Achilles heel of proteomics.** *Nature Biotechnol* 2003, **21:**221–222.

31. McPherson A. ***Crystallization of biological macromolecules.*** Cold Spring Harbor, NY: Cold Spring Harbor Laboratory Press, 1999.

32. Luecke H, Schobert B, Richter H-T, Cartailler J-P, and Lanyi JK. **Structural changes in bacteriorhodopsin during ion transport at 2 Angstrom resolution**. *Science* 1999, **286:**255–260.

33. Lamers MH, Perrakis A, Enzlin JH, Winterwerp HH, de Wind N, and Sixma TK. **The crystal structure of DNA mismatch repair protein MutS binding to a G × T mismatch.** *Nature* 2000, **407**:711–717.

34. Klaholz BP and Moras D. **Structural role of a detergent molecule in retinoic acid nuclear receptor crystals**. *Acta Crystallogr* 2000, **D56**:933–935.

35. Verdino P, Westritschnig K, Valenta R, and Keller W. **The cross-reactive calcium binding pollen allergen, Phl p7, reveals a novel dimer assembly**. *EMBO J* 2002, **21**:5007–5016.

36. Rupp B and Wang J. **Predictive models for protein crystallization.** *Methods* 2004, **34**:390–407.

37. Jurisica I, Rogers P, Glasgow JI, Fortier S, Luft JR, Wolfley JR, Bianca MA, Weeks DR, and DeTitta GT. **Intelligent decision support for protein crystal growth**. *IBM Systems Journal* 2001, **402**:248–264.

38. Han J and Kamber M. ***Data Mining: Concepts and Techniques.*** San Francisco: Morgan Kaufmann Publishers, 2001.

39. Keating S. **LIMS grows from homemade to custom-made solutions**. *Drug Discovery and Development* 2004, **5**:53–54

40. Haebel PW, Arcus VL, Baker EN, and Metcalf P. **LISA: An intranet-based flexible database for protein crystallography project management**. *Acta Crystallog*r 2001, **D57**:1341–1343.

41. Harris M and Jones TA. **Xtrack—A web-based crystallographic notebook**. *Acta Crystallogr* 2002, **D58**:1889–1891

42. Manjasetty BA, Hoppner K, Mueller U, and Heinemann U. **Secure web book to store structural genomics research data**. *J Struct Funct Genomics* 2003, **4**:121–127.

43. Goh CS, Lan N, Echols N, Douglas SM, Milburn D, Bertone P, Xiao R, Ma LC, Zheng D, Wunderlich Z, Acton T, Montelione GT, and Gerstein M. **SPINE 2: A system for collaborative structural proteomics within a federated database framework.** *Nucleic Acids Res* 2003, **31**:2833–2838.

44. Pavlis R. **Top five myths about LIMS interfacing**. *Sci Computing Instrumentation* 2004, **21**:18–19.

45. Goh CS, Lan N, Douglas SM, Wu B, Echols N, Smith A, Milburn D, Montelione GT, Zhao H, and Gerstein M. **Mining the structural genomics pipeline: Identification of protein properties that affect high-throughput experimental analysis**. *J Mol Biol* 2004, **336**:115–130.

46. Bard J and Rhee S. **Ontologies in biology: Design, applications and future challenges**. *Nat Biotechnol* 2004, **5**:213–222.

47. Jancarik J and Kim S-H. **Sparse matrix sampling: A screening method for the crystallization of macromolecules**. *J Appl Cryst* 1991, **24**:409–411.

48. Gilliland GL, Tung M, Blakeslee DM, and Ladner J. **The Biological Macromolecule Crystallization Database, Version 3.0. New features, data, and the NASA archive for protein crystal growth data**. *Acta Crystallogr* 1994, **D50:**408–413.

49. Berman HM, Westbrook J, Feng Z, Gilliland G, Bhat TN, Weissig H, Shindyalov IN, and Bourne PE. **The Protein Data Bank**. *Nucleic Acids Res* 2000, **28**:235–242.

50. Hennessy D, Buchanan B, Subramanian D, Wilkosz PA, and Rosenberg JM. **Statistical methods for the objective design for screening procedures for macromolecular crystallization.** *Acta Crystallogr* 2000, **D56**:817–827.

51. Farr RGJ, Perrymann AL, and Samudzi CT. **Re-clustering the database for crystallization of macromolecules**. *J Cryst Growth* 1998, **183**:653–668.

52. Drenth J and, Haas C. **Nucleation in protein crystallization**. *Acta Crystallogr* 1998, **D54**:867–872.

53. McPherson A, Kuznetsov YG, Malkin A, and Plomp M. **Macromolecular crystal growth as revealed by atomic force microscopy**. *J Struct Biol* 2003, **142**:32–46.

54. Bodenstaff ER, Hoedemaker FJ, Kuil EM, deVrind HPM, and Abrahams JP. **The prospects of nanocrystallography**. *Acta Crystallogr* 2002, **D59**:1901–1906.

55. Drenth J and Haas C. **Protein crystals and their stability.** *J Cryst Growth* 1992, **122**:107–109.

56. Carugo O and Argos P. **Protein-protein crystal-packing contacts**. *Prot Sci* 1997, **6**:2261–2263.

57. Santarsiero BD, Yegian DT, Lee CC, Spraggon G, Gu J, Scheibe D, Uber DC, Cornell EW, Nordmeyer RA, Kolbe WF, Jin J, Jones AL, Jaklevic JM, Schultz PG, and Stevens RC. **An approach to rapid protein crystallization using nanodroplets.** *J Appl Cryst* 2002, **35**:278–281.

58. DeLucas LJ, Bray TL, Nagy L, McCombs D, Chernov N, Hamrick D, Cosenza L, Belgovskiy A, Stoops B, and Chait A. **Efficient protein crystallization.** *J Struct Biol* 2003, **142**:188–206.

59. D'Arcy A. **Crystallizing proteins—A rational approach?** *Acta Crystallogr* 1994, **D50:**469–475.

60. Luft JR, Collins RJ, Fehrman NA, Lauricella AM, Veatch CK, and DeTitta GT. **A deliberate approach to screening for initial crystallization conditions of biological macromolecules.** *J Struct Biol* 2003, **142**:170–179.

61. Chayen NE. **Comparative studies of protein crystallization by vapour-diffusion and microbatch techniques**. *Acta Crystallogr* 1998, **D54**:8–15.

62. D'Arcy A, MacSweeney A, Stihle M, and Haber A. **The advantages of using a modified microbatch method for rapid screening of protein crystallization conditions.** *Acta Crystallogr* 2003, **D59**:396–399.

63. Hansen CL, Skordalakes E, Berger JM, and Quake SR. **A robust and scalable microfluidic metering method that allows protein crystal growth by free interface diffusion.** *Proc Natl Acad Sci USA* 2002, **99**:16531–16536.

64. van der Woerd M, Ferree D, and Pusey M. **The promise of macromolecular crystallization in microfluidic chips.** *J Struct Biol* 2003, **142**:180–187.

65. Baldock P, Mills V, and Stewart P. **Acomparison of microbatch and vapor diffusion for initial screening of crystallization conditions.** *J Cryst Growth* 1996, **168**:170–174.

66. Rupp B. **Maximum-likelihood crystallization.** *J Struct Biol* 2003, **142**:162–169.

67. Chayen N and Saridakis E. **Protein crystallization for genomics: Towards high-throughput optimization techniques.** *Acta Crystallogr* 2002, **D58**:921–927.

68. Chayen NE. **Protein crystallization for genomics: Throughput versus output.** *J Struct Funct Genomics* 2003, **4**:115–120.

69. Segelke BW. **Macromolecular crystallization with microfluidic free-interface diffusion.** *Expert Rev Proteomics* 2005, **2**:165–172.

70. McPherson A. ***Preparation and analysis of protein crystals.*** Malabar, FL: Krieger Publishing Company; 1982.

71. Stura EA, Nemerow GR, and Wilson IA. **Strategies in the crystallization of glycoproteins and protein complexes.** *J Cryst Growth* 1992, **122**:273–285.

72. Segelke BW. **Efficiency analysis of sampling protocols used in protein crystallization screening.** *J Cryst Growth* 2001, **232**:553–562.

73. Carter CW Jr. **Experimental design, quantitative analysis, and the cartography of crystal growth.** In: ***Crystallization of Nucleic Acids and Proteins.*** Ducruix A and Giege R, eds. New York, NY: Oxford University Press, Inc, 1999.

74. Zedzik J and Norinder U. **Statistical analysis and modelling of crystallization outcomes.** *J Appl Cryst* 1997, **30**:502–506.

75. Stewart PS and Baldock P. **Practical experimental design techniques for automatic and manual protein crystallization.** *J Cryst Growth* 1999, **196**:665–673.

76. Prater B, Tuller S, and Wilson L. **Simplex optimization of protein crystallization conditions.** *J Cryst Growth* 1999, **196**:674–684.

77. Tye H. **Application of statistical 'design of experiments' methods in drug discovery.** *Drug Discovery Today* 2004, **9**:485–491.

78. Carter CW Jr and Carter CW. **Protein crystallization using incomplete factorial experiments.** *J Biol Chem* 1979, **254**:12219–12226.

79. Rupp B, Segelke BW, Krupka HI, Lekin TP, Schafer J, Zemla A, Toppani D, Snell G, and Earnest TE. **The TB structural genomics consortium crystallization facility: Towards automation from protein to electron density**. *Acta Crystallogr* 2002, **D58**:1514–1518.

80. Segelke B and Rupp B. **Beyond the sparse matrix screen: A web service for randomly generating crystallization experiments**. *ACA Meeting Series* 1998, **25**:78.

81. Cudney R, Patel S, Weisgraber K, Newhouse Y, and McPherson A. **Screening and optimization strategies for macromolecular crystal growth**. *Acta Crystallogr* 1994, **D50**:414–423.

82. Sivia DS. ***Data Analysis—A Bayesian Tutorial***. Oxford, United Kingdom: Oxford University Press, 1996.

83. Dunlop KV and Hazes B. When less is more: **A more efficient vapour-diffusion protocol**. *Acta Crystallogr* 2003, **D59**:1797–1800.

84. Brzozowski AM and Walton J. **Clear strategy screens for macromolecular crystallization**. *J Appl Crystallogr* 2001, **34**:97–101.

85. Hosfield D, Palan J, Hilgers M, Scheibe D, McRee DE, and Stevens RC. **A fully integrated protein crystallization platform for small-molecule drug discovery.** *J Struct Biol* 2003, **142**:207–217.

86. Page R, Grzechnik SK, Canaves JM, Spraggon G, Kreusch A, Kuhn P, Stevens RC, and Lesley SA. **Shotgun crystallization strategy for structural genomics: An optimized two-tiered crystallization screen against the *Thermotoga maritima* proteome**. *Acta Crystallogr* 2003, **D59**:1028–1037.

87. McPherson A Jr. **Crystallization of proteins from polyethylene glycol**. *J Biol Chem* 1976, **251**:6300–6303.

88. Brzozowski AM and Tolley SP. **Poly(ethylene) glycol monomethyl ethers—An alternative to poly(ethylene) glycols in protein crystallization**. *Acta Crystallogr* 1994, **D50:**466–468.

89. Chayen NE. **Tackling the bottleneck of protein crystallization in the post-genomic era.** *Trends in Biotechnology* 2002, **20**:98.

90. Hillier FS and Lieberman GJ. ***Introduction to Operations Research.*** New York, NY: McGraw-Hill, 2000.

91. McPherson A. **Protein crystallization in the structural genomics era**. *J Struct* Func Genomics 2004, **5**:3–12.

92. DiDonato M, Deacon AM, Klock HE, McMullan D, and Lesley SA. **A scaleable and integrated crystallization pipeline applied to mining the *Thermotoga maritima* proteome.** *J Struct Func Genomics* 2004, **5:**133–146.

93. Koppal T. **Advancing *in vitro* ADME/Tox**. *Drug Discovery and Development* 2004, **5**:47–50.

94. Rupp B. **A guide to automation and data handling in protein crystallization.** In*: Crystallization Strategies for Structural Genomics*. Chayen N, ed. La Jolla–Lisbon: International University Line, 2007.

95. Kim Y, Dementieva I, Zhou M, Wu R, Lezondra L, Quartey P, Joachimiak G, Korolev O, Li H, and Joachimiak A. **Automation of protein purification for structural genomics.** *J Struct Func Genomics* 2004, **5**:111–118.

96. Carter CW Jr and Carter CW. **Protein crystallization using incomplete factorial experiments**. *J Biol Chem* 1979, **254**:12219–12226.

97. Thorne RE, Stum Z, Kmetko J, O'Neill K, and Gillilan R. **Microfabricated mounts for high-throughput macromolecular cryocrystallography.** *J Appl Crystallogr* 2003, **36**:1455–1460.

98. Sanjo A and Cacheu RE. **New microfabricated device technologies for high throughput and high quality protein crystallization**. *ICCBM9 Book of Abstracts* 2002:O.I.2.

99. Cumbaa CA, Lauricella A, Fehrman N, Veatch C, Collins R, Luft J, DeTitta G, and Jurisica I. **Automatic classification of sub-microlitre protein-crystallization trials in 1536-well plates**. *Acta Crystallogr D Biol Crystallog*r 2003, **59**:1619–27.

100. Krupka HI, Rupp B, Segelke BW, Lekin TP, Wright D, Wu H-C, Todd P, and Azarani A. **The high-speed Hydra-Plus-One system for automated high-throughput protein crystallography.** *Acta Crystallogr* 2002, **D58**:1523–1526.

101. Wilson J. **Towards the automated evaluation of crystallization trials.** *Acta Crystallogr* 2002, **D58**:1907–1914.

102. Spraggon G, Lesley SA, Kreusch A, and Priestle JP. **Computational analysis of crystallization trials**. *Acta Crystallogr* 2002, **D58**:1915–1923.

103. Bern M, Goldberg D, Stevens RC, and Kuhn P. **Automatic classification of protein crystallization images using a curve-tracking algorithm**. *J Appl Cryst* 2004, **37**:279–287.

104. Hanson LB, Schall CA, and Bunick GJ. **New techniques in macromolecular cryocrystallography: Macromolecular crystal annealing and cryogenic helium**. *J Struct Biol* 2003, **142**:77–87.

105. Hanson LB, Harp JM, and Bunick GJ. **The well-tempered protein crystal: Annealing macromolecular crystals**. *Meth Enzymol* 2003, **368**:217–235.

106. Garman EF and Doublie S. **Cryocooling of macromolecular crystals: Optimization methods**. *Meth Enzymol* 2003, **368**:188–216.

107. Karain WI, Bourenkov GP, Blume H, and Bartunik HD. **Automated mounting, centering and screening of crystals for high-throughput protein crystallography.** *Acta Crystallogr* 2002, **D58**:1519–1522.

108. Muchmore SW, Olson J, Jones R, Pan J, Blum M, Greer J, Merrick SM, Magdalinos P, and Nienaber VL. **Automated crystal mounting and data collection for protein crystallography.** *Structure* 2000, **8**:243–246.

109. Snell G, Cork C, Nordmeyer R, Cornell E, Meigs G, Yegian D, Jaklevic J, Jin J, Stevens RC, and Earnest TE. **Automatic sample mounting and alignment system for biological crystallography at a synchrotron source.** *Structure* 2004, **12**:1–12.

110. Jacquamet L, Ohana J, Joly J, Legrand P, Kahn R, Borel F, Pirocchi M, Charrault P, Carpentier P, and Ferrer J-L. **A new highly integrated sample environment for protein crystallography.** *Acta Crystallogr* 2004, **D60**:888–894.

111. Viola R, Carmen P, Walsh J, Miller E, and Rupp B. **Pursuing the final frontier in automated high-throughput crystallography: Robotic crystal harvesting.** 5th NIH Bottleneck in Structural Genomics Meeting, NIH/NIGMS, Bethesda, MD, April 13–14, 2006.

3

Automated Liquid Handling Systems for Submicroliter Crystallization

Terese Bergfors

Abstract

A wide range of automated systems for submicroliter crystallization is available nowadays. This chapter traces developments in automation for crystallization and discusses some of the issues in choosing a crystallization robot. To facilitate comparison of the robots, product descriptions are included as an appendix.

3

Automated Liquid Handling Systems for Submicroliter Crystallization

Terese Bergfors

Uppsala University, Uppsala, Sweden

3.1. Introduction

3.1.1. The Development Path of Crystallization Robotics

Setting up crystallization experiments manually is a task inherently fraught with tedium and the risks for random errors, e.g., pipetting into the wrong well. The first commercial crystallization robots to appear in the 1980s, sought to improve the reproducibility and accuracy of crystallization experiments by eliminating as much manual intervention as possible. They were designed to mimic, rather than miniaturize, the manual steps in vapor-diffusion crystallization experiments and most of them used 24-well plates and hanging drops on coverslips.[1-3] During this period, many laboratories also assembled in-house robots in their workshops.

A robot launched in 1990 (Impax, Douglas Instruments, Ltd., www.douglas.co.uk) took a different approach: instead of hanging-drop vapor-diffusion, it automated the microbatch method.[4] This represented a new trend in automation of crystallization because it also miniaturized the experimental setup. Batch crystallization was one of the earliest crystallization techniques but had fallen out of use because it required relatively large amounts of protein compared to the vapor-diffusion method. However, an innovation that incorporated of a layer of oil to prevent evaporation meant much smaller drops could be used.[5] Appropriately, the technique was renamed microbatch. Technically simpler to implement than vapor diffusion,* microbatch lends itself well to automation and all of the crystallization robots nowadays include it in their performance repertoire. For similar reasons, i.e., ease of automation, robot and plate manufacturers began to replace hanging drops on coverslips with sitting drops in specialized plates as a vapor-diffusion technique.

3.1.2. The Shift Towards Nanoliter Crystallization

The 1990s witnessed developments in liquid handling technology that made dispensing of submicroliter volumes feasible. This had a significant impact on crystallization screening. Genomics, proteomics, and a host of other research areas represented potential markets for nanodispensing because of their need to perform many experiments in parallel while simultaneously miniaturizing the scale to conserve materials and reduce costs.

Since it is impossible to manually produce nanoliter drops reproducibly, automation is mandatory. However, there are process considerations, specific to protein crystallization screening, which make a successful adaptation of the nanodispensing technology difficult. Protein crystallization solutions (mother liquors), unlike the liquids in other applications, differ greatly from each other with respect to their viscosity, vapor pressure, and surface tension. The same high pressure necessary to dispense a viscous solution like 30% polyethylene glycol (PEG) causes 20% isopropanol to splatter. In addition, minituarization at the nanoliter level in itself affects the viscosity of liquids during their delivery. Many mother liquors contain highly concentrated salt solutions that can precipitate in the dispensing head and clog the tips. Avoiding cross-contamination between samples is also critical.

---

* Robotic vapor-diffusion experiments require two steps more than the equivalent microbatch setups: the vapor-diffusion reservoirs must be filled with mother liquors and the plates sealed with tape.

3.1.3. Advantages of Crystallization on the Nanoliter Level

Despite the problems in liquid handling that nanoliter dispensing poses (at least for the robot manufacturers), nanocrystallization is extremely attractive. Miniaturization of the crystallization trials allows for more comprehensive screening of the crystallization space with the same or less amount of protein sample. Since this is the most expensive material in the process operation, considerable cost savings can be realized. While simultaneously increasing the number of trials possible, automation also reduces the time required to set up the experiments and improves reproducibility compared to manual setups.

Downsizing of the droplets seems to produce the crystals faster if the protein does crystallize. This is especially beneficial for proteins sensitive to degradation but it also means that the results of screening experiments are obtained more rapidly. For example, four test proteins, which usually crystallized by vapor diffusion in 2–4 days in conventionally sized drops (500–2000 nL) were tested in nanodrops (20–200 nL). For three of them, this reduced their crystallization time from days to hours.[6,7] The same phenomenon has been reported for free-interface diffusion in a microfluidics chip[8] where crystals sometimes grew in only three hours. Along these same lines, experimental designs to accelerate the screening results in microbatch have also been reported, but here it was the type of oil, rather than drop size, that was manipulated.[9]

Nanocrystallization, in addition, may produce crystals that are better ordered since gravity-driven convection is reduced.[10] It may also find crystal forms that are not produced in larger-sized drops.[8,10]

One exciting speculation is that proteins, which do not crystallize in microliter drops, might do so under the different kinetic regime imposed by nanodrops. Some anecdotal reports support this hypothesis; a rigorous investigation will be required to confirm if these observations are valid for macromolecular crystallization in general.

Finally, it can be mentioned that nanodrops are simpler to focus in automated imaging systems because they do not have as much variation in depth-of-focus as larger drops.

3.1.4. Future Robot Technology for Crystallization

The technology in the currently available commercial robots limits the delivery of the drop size to around 20 nL and in practice the drops deployed are usually at least 100 nL, i.e., 50 nL of protein and 50 nL of mother liquor. Reducing the volumes to picoliter level will require a different delivery technology and another type of matrix or substrate than the polystyrene plates in use today. Novel matrices are being fabricated and investigated by several groups for their suitability as high-throughput screening devices at the picoliter level.[11,12] For a discussion of the differences when crystallizing in picoliters rather than nanoliters, see Kuil et al.[13]

Manufacturers of the current generation of robots have focused on automation of vapor diffusion and microbatch. Other methods of exploring the crystallization space, such as free-interface diffusion (FID), liquid-liquid diffusion, and dialysis, are only more recently being automated for high-throughput operations. The commercial platform that has come farthest in this respect, is Fluidigm's TOPAZ™ System (www.fluidigm.com). The TOPAZ™ screening chip is a polydimethylsiloxane elastomer with microfluidic channels for performing FID. With 1.5 µL of protein solution, 96 conditions can be screened. The original chip prototype has been reformatted with an SBS (Society for Biomolecular Screening) footprint so that protein and reagents can be added to the inlets by a robot or hand-held multichannel pipette. It takes approximately five minutes to load the chip after which pressure is applied to drive the fluids into the microchannels. When the containment valves are then opened, the protein and mother liquor solutions begin to diffuse into each other. The system can be expanded to include a workstation for inspection of the chips. A second chip design, for optimization, can be mounted directly on a goniometer for inspection of the crystals in the X-ray beam.

Adaptation of the dialysis method for high-throughput crystallization is also being investigated. Dr. Stephen Lee, University of Oxford, has designed a device to adapt the 96-well crystallization plates for high-throughput dialysis. This is currently under development by ISIS Innovation, Ltd. (www.isis-innovation.com).

3.2. Choosing a Robot

3.2.1. Background of This Chapter

As an academic research lab, our interest in crystallization robotics is for the advantages that it offers in screening scarce amounts of valuable proteins rather than for its role in a pipeline production. The product profiles presented in this chapter are the result of our market investigation when we in 2005 received funding for the instrumentation. For any crystallographic laboratory looking to improve its efficiency through automation, a crystallization robot is arguably the most essential piece of equipment to buy so that is why we focus on this particular aspect of automation here. More laboratory automation for high-throughput proteomics is described elsewhere in this book and in recent reviews by Bard et al.,[14] Hui and Edwards,[15] Kuhn et al.,[16], and Stevens.[7]

3.2.2. What are the Automation Requirements?

With the variety of commercial systems now available, laboratories should be able to find a system that matches their specific automation requirements. In deciding which robot might be most appropriate for the experimenter's requirements, the following points are worth consideration.

3.2.3. Multiuser or Service Laboratory?

How many users will be manipulating the instrument? If there are not specific personnel dedicated to the robot, maintenance is an important consideration. Daily and weekly washing routines are mandatory for models that do not use disposable tips. How robust is the robot? The risks of breakage increase in multiuser laboratories.

The software must be especially user-friendly when the researchers will be running their own samples rather than submitting them to a dedicated service unit. It is advantageous if the software can be reprogrammed to accommodate new crystallization plate footprints when they appear. The software and robot hardware should be compatible with workstations performing functions related to handling of the crystallization plates, e.g., plate loaders, bar coders, storage hotels, retrieval for image analysis, and so on.

3.2.4. Reservoir Filling and Cocktail Production

Another important decision in choosing the crystallization robot is whether it should be single or multipurpose, **Table 3.1**. The single-function robots produce the composite protein/mother-liquor droplet. This is the only dispensing step necessary for microbatch experiments but in vapor-diffusion setups, the mother liquors must also be dispensed into the reservoirs of the plate.* The volume ranges for these two dispensing operations are quite different, e.g., 50–100 nL for the protein/mother liquor droplets vs. 100 μL for reservoirs. Dual-function machines will therefore be complicated instruments since they require different dispensing syringes, mechanisms, or technologies if one dispensing range it not to be compromised at the expense of the other. Yet another consideration when the same robot is used to prepare both the nanodrops and fill the vapor-diffusion plate reservoirs is that the latter can become a rate-limiting step. So, despite the extra expense and maintenance that two handlers entail, many laboratories choose to have separate robots for dispensing mother liquors into the reservoirs and making the composite nanodroplets of protein/mother liquor.

The initial screening step in crystallization often uses fixed mixtures of mother liquors. However, at the optimization level of crystallization work, the mother liquors must always be generated in new combinations, i.e., cocktail production is required. Several of the crystallization robots in **Table 3.1** create screens, at least in some limited grid fashion. Other liquid handlers, like the Alchemist I™ (RoboDesign International, Inc., www.robodesign.com) and Matrix Maker™ (Emerald BioSystems, Inc., www.emeraldbiostructures.com), are dedicated solely to cocktail production and are intended for incorporation with the dispensing robotics.

---

* The mother liquors must be aspirated from some bulk container. Some robots can do this only from SBS-formatted plates, e.g., Deep Well blocks and microtiter plates but others can accept even non-SBS formats such as 15 mL polystyrene tubes and 100 mL Pyrex bottles.

Table 3.1. Liquid handling capabilities of the different crystallization robots.

Manufacturer or Distributor URLs	Robot series	Dispensing of the protein/mother liquor droplets	Fills the vapor diffusion plate reservoirs	Cocktail production/screen formulation
Art Robbins Instruments, Inc. www.artrobbinsinstruments.com	Phoenix™	SD, M	Yes	No
Cy-Bio AG www.cybio-ag.com	CyBi® -HTPC	SD, M	Yes	No
Deerac Fluidics, Ltd. (formerly Allegro Technologies) www.deerac.com	Equator™ NS101, NS808	SD, M	No	No
Douglas Instruments, Ltd. www.douglasinstruments.com	Oryx 6™	SD, M	No	Yes
Emerald BioSystems, Inc. www.emeraldbiosystems.com	Drop Maker™ Matrix Maker™ Matrix Drop Maker Hybrid™	SD, M No SD, M	No Yes Yes	No Yes Yes
Genomics Solutions, Ltd. www.genomicsolutions.co.uk www.cartesiantech.co	MicroSys 8™ Honeybee 161™	SD, M SD, M	No No	No No
Gilson, Inc. www.gilson.com	Constellation® 1200	SD, M	Yes	No
Hamilton Bonaduz AG www.hamiltoncomp.com	Microlab® Star	SD, M	Yes	Yes
Innovadyne Technologies, Inc. www.innovadyne.com	Screenmaker™ 96+8 Nanodrop™ II, Nanodrop™, ExtY	SD, M SD, M SD, M	Yes Yes No	Yes Yes No
Matrix Technologies Corporation, Ltd. www.matrixtechcorp.com	Hydra® II Plus-One	SD, M	Yes	No
RoboDesign, Inc. www.robodesign.com	Alchemist I™ Kalypso™	No SD, M	No Yes	Yes No
Tecan Group, Ltd. www.tecan.com	Freedom EVO®	SD, M	Yes	Yes
TTP LabTech, Ltd. www.ttplabtech.com Molecular Dimensions, Ltd. www.moleculardimensions.com	Mosquito™	SD, M	No	No

Abbreviations: SD = sitting drop, M = microbatch.
Note: Most of the robots that can deliver sitting drops are also capable of producing hanging drops in some fashion, although user intervention is required, e.g., to flip the hanging-drop supports. The Freedom EVO® (Tecan) is the only robot with fully automated hanging drop production. Detailed product descriptions for specific robots can be found in the **Appendix 3.1** of this chapter.

3.2.5. Types of Dispensing

There are two main types of delivery mechanisms in robotic pipetting: contact and noncontact, **Table 3.2**. In contact dispensing, the liquid is dispensed either by air displacement (an air cushion pushes the liquid out) or positive displacement (a piston forces the fluid through the orifice). Both of these methods have been used for many years in handheld pipettes and are also used in crystallization robots. A third method

Table 3.2. Different liquid handling technologies in contact and noncontact dispensing.

Contact dispensing	Non-contact dispensing
Air displacement 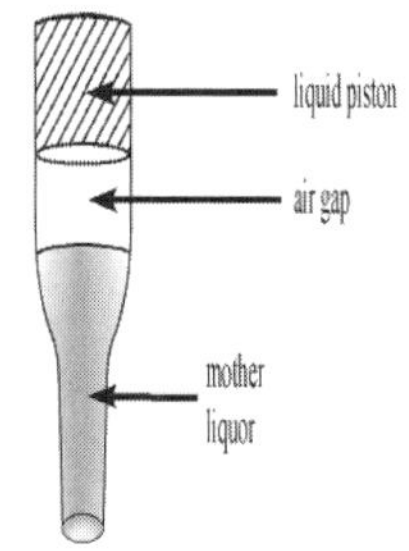	*Syringe pump coupled to solenoid valve* 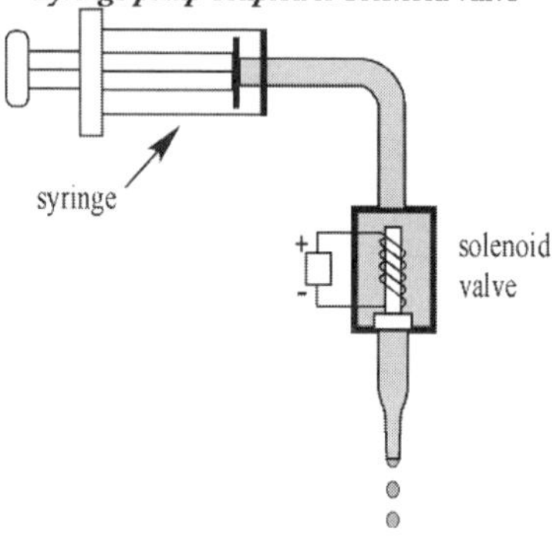
Positive displacement 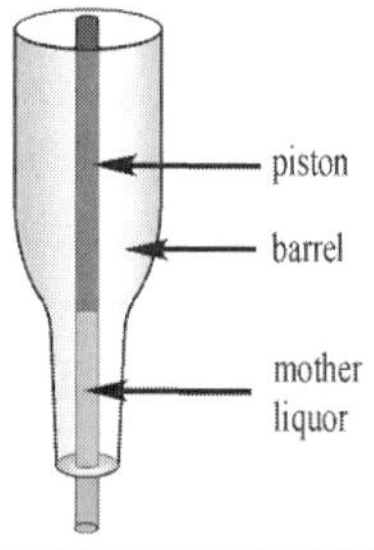	*Piezoelectric* 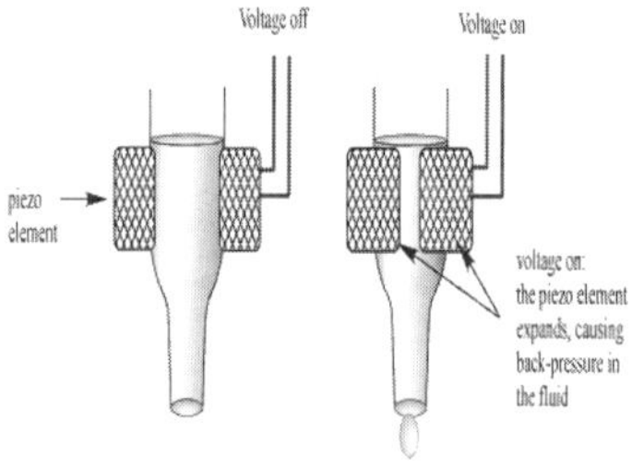
Pin tool 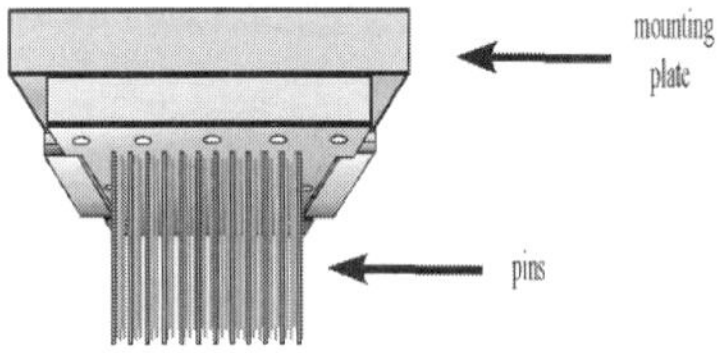	*Electrospray* 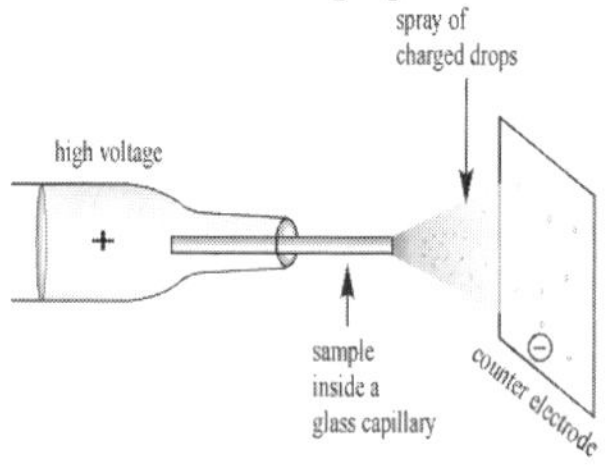

of contact dispensing, the pin tool, operates by dipping the pins into the liquid to be transferred. The liquid adheres to the sides of the pins, which may also contain slots. Pin tools are used, for example, in "cherry-picking" from compound libraries. This particular technology is not deployed in crystallization robots, however.

Ink-jet technology (so called because of its origins in the printing industry) is a noncontact dispensing method and many of the crystallization robots use some variation of it. Essentially it operates by creating backpressure in the fluid to be delivered. There are various ways for doing this, e.g., with a syringe pump coupled to a microsolenoid valve, or by applying voltage to a piezoelectric transducer.* When the fluid's container is opened, the liquid is ballistically expelled due to the accumulated backpressure. The flow of liquid can be modulated to create droplets of varying size by manipulation of the backpressure level, the diameter of the orifice, the opening duration, or some combination thereof.

Electrospray, e.g., in mass-spectroscopy applications, is another kind of noncontact dispensing, which operates by ionization of the liquid. It is not used in any commercial crystallization robots yet, but at least one prototype is underway.[10]

3.2.5.1. Contact or Noncontact Dispensing: Does It Matter?

No one particular type of dispensing mode is without its respective advantages and disadvantages: The variety of liquid-delivery technologies in the different robots presented here is partially a reflection of this. It is difficult to say (although robot manufacturers might argue otherwise) if one type of dispensing is inherently superior to the others since there are so many different process considerations in handling the liquids typically used in protein crystallization.

One question that has been posed is how different delivery methods affect the mixing of the protein and mother liquor. Mixing increases nucleation and therefore, at least in theory, should potentially generate more leads during screening. Most of the robots deliver the droplets of mother liquor and protein sequentially; the software usually includes a stirring option for mixing the two components after they are deposited on the plate. One exception is the Oryx (Douglas Instruments, Ltd., www.douglas.co.uk): The mother liquor and the protein are aspirated into

* Thermal ink-jet devices actually vaporize the liquid into a bubble when current is applied to the liquid, usually ink. For obvious reasons, this thermal/bubble technique is not relevant within the context of protein crystallization.

the same dispensing tip (although in different bores) and dispensed simultaneously as a single drop, i.e., mixing occurs as the drop is delivered to the plate. Additional stirring afterwards in the plate is also possible.

The mixing effects of sequential vs. simultaneous delivery and contact vs. noncontact dispensing are only beginning to be investigated. It is an expensive evaluation to make, requiring the purchase of several robots for the purpose of comparison. It has been speculated that there is more instantaneous mixing of the protein and mother liquor components in noncontact dispensing because of the velocity at which the second droplet hits the first droplet but this remains to be confirmed.[14]

3.2.6. Accuracy and Precision

The product descriptions of the different robots at the end of this chapter include the percent coefficient of variation, (%CV), where available, for the delivery of a high concentration of PEG. PEG is often chosen as a representative example because of its viscosity; the user might also compare the manufacturers' %CVs for mother liquors with organic solvents and salts.*

The %CV is a measure of the precision of the delivery, i.e., reproducibility. It is defined as:

$$\%CV = \frac{\text{standard deviation of the dispenses}}{\text{mean value of the dispenses}} \times 100$$

The %CV can be determined for either a dry or wet transfer; this is not always specified by the manufacturer. A dry transfer, i.e., touch-off to a dry surface, is the more difficult to achieve.

In high-throughput operations, reproducibility (precision) is more important than the absolute values delivered (accuracy). For this reason, it is usually the %CVs that are reported in technical specifications. Accuracy, on the other hand, is reported as percent of bias and it measures what the actual volume is compared to what is set to be delivered on the robot (or pipette's) display. For example, if 95 nL were delivered instead of the expected 100 nL, the accuracy error would be 5%.

---

* One solution, which is notoriously difficult to dispense is G11 in the Crystal Screen HT (Hampton Research, Inc., www.hamptonresearch.com). It contains 70% methyl-2,4-pentanediol (MPD), i.e., it has both low surface tension and high viscosity, (Janet Newman, personal communication.)

Here we have been referring to accuracy and precision with respect to delivery of the liquid volumes. Geometric accuracy and precision, i.e., the *x-y-z* positioning of the drops on the substrate (usually an SBS-formatted polystyrene crystallization plate), are also important considerations. These parameters are adjusted by the manufacturers upon installation of the robot and then an alignment jig ensures correct positioning of the plate each time it is loaded, either by an automated plate loader or manually by the user.

3.2.7. Speed and Related Issues of Chilling and Evaporation Control

The speed at which the robots set up the composite droplet of protein and mother liquor is an essential issue because of the vulnerability of such small drop volumes to evaporation. Some systems are so rapid (under 2 min) that evaporation in the droplets during the course of the dispensing is considered negligible. Slower systems include some kind of shielding device or humidifying chamber that may be passive or employed at the user's discretion. Vapor-diffusion experiments must then be sealed when the plate is completed. If there is no automatic plate sealer incorporated in the platform, this task becomes the responsibility of the operator. Most people can supervise a plate attentively for short periods like two minutes. Longer dispensing times, however, increase the risks for disruption of the user's attention and possibly jeopardize the sealing of the plate.

In microbatch experiments, the oil layer effectively eliminates evaporation during the course of the dispensing and the need for a plate sealer afterwards. Depending on the robot, the mother liquor and protein droplets can be added to pre-oiled dishes. However, if the droplets are dispensed to dry plates, which are subsequently filled with the oil, then the speed of delivery becomes an issue in microbatch as well.

Chilling, either with Peltier elements or recirculating chillers, is available on some robots for the cassette that holds the protein sample(s). More advanced models will also chill the plate locations or mother liquor stocks. Of course, a humidifying chamber cannot be deployed in combination with chilling of the destination plate because this would cause condensation.

For pipeline operations, the true turnaround time must also be considered, i.e., the time before the next plate can be loaded onto the platform and processed after washing cycles, changing of tips, sealing of the previous plate, etc.

3.2.8. Size

Even for spacious laboratories this can be an issue if the instrument needs to be placed inside an anaerobic chamber or moved into a temperature-controlled cabinet. Note that some robots require a source of water or helium for their operation.

3.2.9. Price

Prices are quickly outdated by market developments and fluctuations in the exchange rates. With that said, the buyer should expect to pay at least 50,000 euros for the cheapest system and more typically 65,000 to 100,000 euros, at a minimum, for the other liquid handling systems. (Prices are current as of 2005.)

This level of expense is not a problem for high-throughput proteomics consortia but for small laboratories that cannot afford a robot, the Nanoject II (Drummond Scientific Co., www.drummondsci.com) is worth considering. This nanoinjector operates by positive displacement to handle variable volumes from 2.3 to 69 nL. The intended application of this instrument is for performing nanoliter injections into living cells but Yeh's group reports the use of it in crystallization screening.[17] This foot-pedal operated pipette offers the possibility of performing nanodrops at an extremely modest price compared to that of a dedicated robot.

Another group has experimented with an ordinary ink-jet printer head to crystallize lysozyme and was able to achieve drop sizes as low as 2 nL.[18]

3.2.10. Product Descriptions

Appendix 3.1 to this chapter presents the different commercial crystallization robots in a table form to facilitate comparisons between them. Pictures of the different systems are found in **Plate 3.1**. Every effort has been made to cover the market comprehensively; any oversights are unintentional. The content was compiled from the manufacturers' technical specifications or supplied by them to ensure the accuracy of the information. However, due to the rapid development in this field, the reader should contact the distributors or manufacturers directly for the most up-to-date descriptions.

Acknowledgements

I thank Dr. Janet Newman and the robot manufacturers for useful discussions and Drs. Matti Nikkola and Elisabeth Sauer-Eriksson for critically reading the manuscript.

References

1. Cox M and Weber P. **Experiments with automated protein crystallization**. *J Appl Cryst* 1987, **20**:366–373.
2. Oldfield T, Ceska T, and Brady R. **A flexible approach to automated protein crystallization**. *J Appl Cryst* 1991, **24:**255–260.
3. Sadaoui N, Janin J, and Lewit-Bentley A. **TAOS: An automatic system for protein crystallization.** *J Appl Cryst* 1994, **27**:622–626.
4. Chayen N and Stewart P**. An automated system for micro-batch protein crystallization and screening**. *J Appl Cryst* 1990, **23**:297–302.
5. Chayen N, Stewart PS, and Blow D. **Microbatch crystalliztion under oil - A new technique allowing many small-volume crystallization trials.** *J Cryst Growth* 1992, **122:**176–180.
6. Santarsiero B, Yegian D, Lee C, Spraggon G, Gu J, Scheibe D, Uber D, Cornell E, Nordmeyer R, Kolbe W, Jin J, Jones A, Jaklevic J, Schultz P, and Stevens R. **An approach to rapid protein crystallization using nanodroplets.** *J Appl Crys* 2002, **35**:278–281.
7. Stevens R. **High-throughput protein crystallization.** *Curr Opin Struct Biol* 2000, **10**:558–563.
8. Hansen C, Skordalakes E, Berger J, and Quake S. **A robust and scalable microfluidic metering method that allows protein crystal growth by free interface diffusion.** *PNAS* 2002, **99**:16531–16536.

9. D'Arcy A, MacSweeney A, Stihle M, and Haber A. **The advantages of using a modified microbatch method for rapid screening of protein crystallization conditions.** *Acta Cryst* 2003, **D59:**396–399.

10. Bodenstaff E, Hoedemaeker F, Kuil M, de Vrind H, and Abrahams J. **The prospects of protein nanocrystallography.** *Acta Cryst* 2002, **D58:**1901–1906.

11. Juarez-Martinez G, Steinmann P, Roszak A, Isaacs N, and Cooper J. **High-throughput screens for postgenomics: studies of protein crystallization using microsystems technology.** *Anal Chem* 2002, **74**:3505–3510.

12. Pechkova E and Nicolini C. **Protein nanocrystallography: A new approach to structural proteomics.** *Trends Biotechnol* 2004, **22**:117–122.

13. Kuil M, Bodenstaff E, Hoedemaeker F, and Abrahams J. **Protein nano-crystallogenesis. Enzyme Microb Technol** 2002, **30:**262-265.

14. Bard J, Ercolani K, Svenson K, Olland A, and Somers W. **Automated systems for protein crystallization.** *Methods* 2004, **34**:329–347.

15. Hui R and Edwards A. **High-throughput protein crystallization.** *J Struct Biol* 2003, **142**:154–161.

16. Kuhn P, Wilson K, Patch M, and Stevens R. **The genesis of high-throughput structure-based drug discovery using protein crystallography**. *Curr Opin Chem Biol* 2002, **6**:704–710.

17. Yeh J. **A manual nanoscale method for protein crystallization.** *Acta Cryst* 2003, **D59**:1408–1413.

18. Howard E and Cachau R. **Ink-jet printer heads for ultra-small-drop protein crystallography.** *BioTechniques* 2002, **33**:1302–1306.

Appendix 3.1.

Product descriptions of the crystallization robots listed in Table 3.1.

(Arranged alphabetically by the manufacturer's name.)

Unless otherwise specified, "speed" refers to the dispensing times for a 96-well, SBS vapor-diffusion plate. The term "96/3-plate" refers to the SBS-plates with 3-indentations, e.g., Greiner's CrystalQuick™ 96. These plates have 288 small-volume chambers for the protein/mother liquor droplets, and 96 mother-liquor reservoirs.

Manufacturer: Art Robbins Instruments, Inc. **Robot:** Phoenix™

Description	The Phoenix™ is a 9-position dispenser with a 96-well dipensing syringe head and an independent single or multichannel for the protein. Developed by the same people that made the Hydra series; it is a fast, low-cost, robust instrument, simple to use and service.
Comments to **Table 3.1**	The Phoenix™ can accommodate Deep Wells and microtiter plates for the mother-liquor transfers. Microbatch experiments can be dispensed through the oil or to dry plates.
Dispensing volumes	100 nL to 290 nL with the 96-well head. 50 nL to 15 μL with the single-channel dispenser.
Method of dispensing	Dispensing of the mother liquors is by positive-displacement syringes (choice of either Teflon-coated stainless steel or flexible Nitinol). The protein is delivered by noncontact dispensing with a single-channel head.
Examples of reported %CVs	Viscosity is not a problem in positive displacement. Overall CVs of 5.1% at 100 nL.
Number of plates that can be loaded "walk-away"	Up to 9.
Non-SBS formats, 24- or 48-well plates accepted?	Yes, with modifications and 48- or 24-syringe head options but the destination-plate footprint must be SBS-format.
Humidity control?	No.
Chilling available?	Yes, for the protein. One or two plates could also be chilled.
Speed	Approximately 2 min, including filling of the reservoirs with mother liquor. Turnaround time for washing is an additional 2 min.
Associated costs, consumables required, breakable items, etc.	If proper washing of the instrument is practiced, the only cost would be replacement of the Teflon tips on the syringes (under US$400 per year). No helium required.
Size	63 × 70 × 71 cm; 64 kg.

Information and picture generously provided by Art Robbins, Art Robbins Instruments, Inc., USA, and Techtum AB (their Swedish distributor).

Manufacturer: CyBio AG **Robot:** CyBi®-HTPC

Description	A dedicated system, which combines a 96-tip pipetting head for transferring the mother liquors and an 8-tip system for the protein solutions.
Comments to **Table 3.1**	Hanging drops can be prepared but the support must be turned manually after pipetting. The oil for microbatch can be pipetted with the system.
Dispensing volumes	Minimal volume: 0.2 μL (dry transfer). Maximum volume in one dispense: 20 μL.
Method of dispensing	Contact dispensing with disposable tips; tips can be washed if necessary. Deionized water is used to wash the tips (no degassing required) is used to wash the tips.
Examples of reported %CVs	Typically less than 5% at 1 μL; less than 15% at 0.2 μL.
Number of plates that can be loaded "walk-away"	Up to 4 in the standard system version. Upgradeable to 40 plates for microbatch methods.
Non-SBS formats, 24- or 48-well plates accepted?	No.
Humidity control?	In the standard version, evaporation problems are overcome by speed.
Chilling available?	Optional.
Speed	Circa 4–5 min for a 96-well plate; this includes the time to fill the reservoirs with mother liquor. The 96/3-well plate takes 9 min. Parallel processing allows the 96-well head to be washed while the 8-tip head aspirates the protein.
Associated costs, consumables day required, breakable items, etc.	One box with 96 tips for the 96-channel head (8 Euros) per and 8 tips per set of experiments for the 8-channel head that delivers the protein.
Size	Available on a stand-alone table (100 × 90 × 70 cm without the computer) or as a desk top unit (120 × 100 × 155 cm).

Information and picture generously provided by Dr. Alexander Frittrang, CyBio AG, Jena, Germany.

Manufacturer: Deerac Fluidics **Robot series:** Equator™ NS101, NS808

Description	The Equators™ are two bench-top pipetting robots that dipense the nanodrops of protein and mother liquor. The Equator 101 is a single-channel and the Equator 808 is an 8-channel.
Comments to **Table 3.1**	The Equators™ are available as stand-alone instruments or for incorporation onto a larger robotic work station, e.g., the Tecan EVO®. The dispensing head is available as a separate "Spot-on™ module" for integration onto a third-party's robotic plat form.
Dispensing volumes	50 nL to 20 μL.
Method of dispensing	Noncontact, Spot-on™ technology, which uses a time-pressure valve integrated into the tip. The proprietary FastWash mechanism ensures similar cleaning to disposable tips. No system liquid (helium, water, etc.) or air gaps required in the dispense mechanism; this ensures minimal dead volume.
Examples of reported %CVs	Noncontact dispense as low as 50 nL with CVs <10%.
Number of plates that can be loaded "walk-away"	With a plate stacker more than 40 plates.
Non-SBS formats, 24- or 48-well plates accepted?	Yes. The software can accommodate custom plate designs with multiple layers and shelves.
Humidity control?	No. New sealing mechanism which greatly minimizes evaporation effects is available (contact: info@deerac.com).
Chilling available?	Yes.
Speed	Dispenses 384 protein drops in 8 s.
Associated costs, consumables required, breakable items, etc.	There is no major associated cost in that dispense tips are fixed. Lifetime can be in excess of 1 million dispenses. Cost of replacement is lower than other fixed tip systems.
Size	67 × 38.8 × 37.3 cm.

Information and picture generously provided by Devin Donnelly, Deerac Fluidics, Ireland.

Manufacturer: Douglas Instruments, Ltd.	**Robot:** Oryx 6™
Description	Oryx 6™ performs screening and optimization crystallization experiments for microbatch or sitting-drop vapor diffusion.
Comments to **Table 3.1**	Screening: uses, e.g., 96 solutions in a microtiter plate. The screening and protein solutions are added simultaneously as the drop is delivered to the destination plate. The wells are individ ually oiled within 5 s of dispens-ing each drop. Drops can be added to pre-oiled plates, but dry transfers are recommended for better accuracy. Can perform microbatch additive screens, combining protein, cocktail, and up to 96 additives. Optimization is possible with up to five solutions, e.g., protein, precipitant, additive, buffer, and water. The X-Step software generates gradient and multi variate experiments for optimization.
Dispensing volumes	100 nL to 10 µL.
Method of dispensing	Contact dispensing with a two-bore tip (made of a fluorinated polymer material) for screening and a 5-bore tip for optimiza tion. Degassed water is used as a system fluid to fill syringes, etc. Air bubbles, usually 1 µL, separate sample from the system fluid. Dead volume for the protein is only 1%.
Examples of reported %CVs	CVs have been measured for the optimization setup using 50% PEG 4000 and 95% PEG 600 (with 5% red dye). 1 µL: CV = 1%; 100 nL: CV = 5% The CV for sitting-drop dispensing (100 nL plus 100 nL) is 7%
Number of plates that can be loaded "walk-away"	1 (2 used occasionally).
Non-SBS formats, 24- or 48-well plates accepted?	Yes. Any plate with dimensions of a Linbro plate or smaller.
Humidity control?	No.
Chilling available?	No.
Speed	Sitting drop transfers: 96 wells in 12 min, including rinses and protein loading. Drops are covered by a sliding cover after 60 s or less. Microbatch optimization: 48 wells in 8 min is a typical result.
Associated costs, consumables required, breakable items, etc.	After several years of intensive use, belts may need to be replaced. Additional bore tips (2–6) might be required; average annual cost is US$60.
Size	67 × 61 cm (excluding the controller and laptop). Complete system including controller = 34 kg.

Information and picture generously provided by Patrick Shaw Stewart, Douglas Instruments, Berkshire, United Kingdom.

Manufacturer: Emerald BioSystems, Inc. **Robot**: Matrix-Drop Maker Hybrid™

Description	There are four principle components in the Matrix-Drop Maker Hybrid™ system: the Matrix Maker™ prepares the mother liquors which the Matrix Transfer Station transfers to the plate reservoirs; the MicroShot Dispense Nozzle dispenses the droplets of protein/mother liquor and the Crystal Miner™ relational-database software stores all trial data.
Comments to **Table 3.1**	The Matrix Maker™ can create mother-liquor cocktails from up to 60 different stock solutions, in small to large volumes, from bulk stocks in conical tubes, Deep Well blocks, or 50- and 100-mL bottles.
Dispensing volumes	Matrix Maker™: from 6 μL to 50 mL. MicroShot Dispense Nozzle: 100 to 500 nL.
Method of dispensing	Stock solutions are dispensed by positive displacement using precision-syringe pumps and stainless-steel needles. The dispense head with 24 MicroShot Dispense nozzles uses noncontact dispensing.
Examples of reported %CVs	Matrix Maker™: ± 2% for both non-viscous and viscous solutions down to 20 μL. MicroShot Dispense Nozzle: Overall CV <7% and ± 10% at 100 nL with 50% PEG 8000.
Number of plates that can be loaded "walk-away"	Matrix-Drop Maker Hybrid™ : 1–9 plates. Matrix Maker™ : 16 SBS plates.
Non-SBS formats, 24- o 48-well plates accepted?	Matrix-Drop Maker Hybrid™ : no. Matrix Maker™ : yes
Humidity control?	No?
Chilling available?	Yes. Optional temperature-controlled deck. Closeable sash included.
Speed	Cocktail production: 96 solutions in less than 40 min. Crystallization trial setup in less than 6 min (including filling of the reservoirs).
Associated costs, consumables required, breakable items, etc.	Pipette tip consumables for transfer of mother-liquor solutions (cocktails) from their source block to the crystallization plate but no consumables for the cocktail production or setting-up of drops. Includes one-year warranty and technical support.
Size	125 × 85 × 178 cm; 295 kg.

Information and picture generously provided by Craig Sterling, Emerald BioSystems, Inc., USA.

Manufacturer: Genomics Solutions, Ltd. **Robot series:** Honeybees 161™ and MicroSys 8™

Description	Genomics Solutions markets two robots for protein crystallization: the Honeybee 161™ (based on the synQUAD™ PixSys with 8 or 16 channels and the possibility of an extra protein channel) and the MicroSys 8™.
Comments to **Table 3.1**	The Honeybee 161™ and MicroSys 8™ dispense the drops; reservoirs must be prefilled. Hanging drops are possible using Greiner's CrystalDrop™; the user must fit the lid manually onto the reservoir plate. Microbatch: Oil is dispensed first by hand or another robot. The fluidic inertia means that the protein and mother liquors are "shot through" the oil. The plate is then centrifuged to make the two solutions coalesce.
Dispensing volumes	50 nL to 20 µL
Method of dispensing	Noncontact. The technology, patented under the name synQUAD™, synchronizes the translation of the destination plate platform with noncontact delivery of the drops by a microsolenoid valve and syringe pump.
Examples of reported %CVs	20–30% PEG 8000: <7.1% for 100-nL transfers (dry transfer).
Number of plates that can be loaded "walk-away"	MicroSys 8™: 3 plates. Honeybee 161™: 9 and up to 161 with the optional plate loader.
Non-SBS formats, 24- or 48-well plates accepted?	Yes.
Humidity control?	Yes.
Chilling available?	Honeybees 161™: yes, for the protein. MicroSys 8™: no.
Speed	For a 96/3-plate (both robots): 10–12 min for dispensing the mother liquors onto the droplet shelves and subsequent washing steps. The protein solution is dispensed on-the-fly (20 s).
Associated costs, consumables required, breakable items, etc.	Valves have a finite life but last a long time with proper maintenance. Tips can break, usually this is user error. Training tips are provided. Degassed water required.
Size	Honeybee 161™ = 120 × 105 × 50 cm MicroSys 8™, with humidity chamber = 127 × 70 × 65 cm.

Information and picture generously provided by Genomics Solutions, Cambridgeshire, United Kingdom and their Swedish distributor, LabVisions AB.

Manufacturer: Gilson, Inc. **Robot:** Constellation® 1200

Description	The Constellation® 1200 is a precise small-liquid handler consisting of a 12-probe Z-arm configuration for applications ranging from MALDI-spotting and protein arrays to crystallography.
Comments to **Table 3.1**	Fills the reservoirs for vapor diffusion from SBS-formats like DeepWell blocks and microtiter plates. Microbatch: the robot dispenses the oils (paraffin, Al's or silicon oils, etc.). Protein solution and mother liquors are "shot-through" the oil with a near-touch-off method. Capable of dispensing with or without an air gap.
Dispensing volumes	Dispenses from 50 nL to 500 µL volumes into 96-, 384-, and 1536-well plates employing solenoids for small volumes and syringes for large volumes. Versions of the instrument are available for higher-density plates and volumes down to 5 nL.
Method of dispensing	Noncontact using syringe-solenoid technology. The syringe fluid pathways are independent of the solenoids, eliminating the possibility of plugging or contamination of the solenoid valves and also increasing solenoid life-time compared to other syringe-solenoid technology.
Examples of reported %CVs	11.1% for 30% PEG 8000 with 100-nL dispenses.
Number of plates that can be loaded "walk-away"	Up to 12 plates.
Non-SBS formats, 24- or 48-well plates accepted?	Yes. 24- and 48-well plates are acceptable as destination plates.
Humidity control?	No.
Chilling available?	The deck can be chilled with an external chiller device such as the Julabo F32-HE chiller.
Speed	Processes one 1536-well plate in under 5 min when dispensing 200 nL or less. The approximate time to transfer 96 mother liquors to a 96-well plate is 5 min and 1 min to transfer the protein sample. A 96/3 plate requires about 13 min.
Associated costs, consumables required, breakable items, etc.	The only parts that typically need to be replaced are two syringes, used for sample aspiration, and the solenoids; (they have a million-pulse life cycle). Uses helium as a pressure supply.
Size	115 × 68 × 76 cm; 120.7 kg.

Information and picture generously provided by Dr. T. Scott Kephart, Gilson, Inc., USA.

Manufacturer: Hamilton Bonaduz AG **Robot series:** Microlab® Star

Description	The Microlab® Star, a contact dispenser, is a member of the STARline family. It uses monitored air-displacement pipetting technology on individual channels; no system tubing is required. The patented CORE (**C**ompression Induced **O**-**R**ing **E**xpansion) technology allows aspiration by disposable tips and needles in parallel with the highest possible dispensing accuracy in *xyz* directions.
Comments to **Table 3.1**	The Microlab® Star can fill plate reservoirs from bulk containers, e.g., Deep Well blocks or 15-mL conical tubes, and create grid screens. Microbatch: the robot dispenses the oil first; then adds the protein and reagents.
Dispensing volumes	1–1000 μL.
Method of dispensing	Contact dispensing: monitored air-displacement. Tips and needles can be used in parallel mode depending on the viscosity of the liquid, which has to be transferred and the mode of dispensing (surface, jet, or aliquoting).
Examples of reported %CVs	<2% at 20 μL 30% PEG 8000. <6 % at 2 μL 30% PEG 8000.
Number of plates that can be loaded "walk-away"	Up 520 plates on deck depending on deck size (Starlet, Star, or StarPlus). Integration of storage units possible.
Non-SBS formats, 24- or 48-well plates accepted?	Yes, all available plates on the market.
Humidity control?	As an option.
Chilling available?	Yes. Chilling is available for the proteins and destination plates.
Speed	Transfer of ready-made mother liquors to plate reservoirs takes 13 min. Dispensing of 96 protein/mother liquor droplets: 30 min. Creation of custom mother liquors from stock solutions takes 20 to 100 min depend-ing on the gradient /mixture complexity. There are three sets of 8 needles: one set is in use while the other two are washing to eliminate waiting.
Associated costs, consumables required, breakable items, etc.	Market-oriented costs of disposable tips. No hardware maintenance required when used as intended. No system fluids or degassed water needed.
Size	Starlet: 113 × 91 × 78 cm; 135 kg. Star: 167 × 91 × 78 cm; 145 kg. StarPlus: 216 × 91 × 78 cm; 205 kg.

Information and picture generously provided by Drs. Carsten Mang and Carsten Cetzold, Hamilton Bonaduz AG, Switzerland.

Manufacturer: Innovadyne Technologies, Inc.

Robot series: ScreenMaker™ 96+8, Nanodrop™ II, ExtY

Description	The Screenmaker™ 96 +8 provides multiple liquid-handling functions on a flexible single instrument. The +8 can add different reagents or protein to the crystallization plates from a Deep Well block or other source plate. Tip independence on the 96-head allows addition of individual reagents to any well, facilitating "on-the-fly" screen creation. The Nanodrop™ Xtal platforms are affordable alternatives to the more powerful Screenmaker™ 96 +8. They are available as single-plate (Nanodrop™ ExtY) or two-plate (Nanodrop™ II) configurations. The two-plate platform can fill reservoirs as well as perform the mother liquor/protein droplets.
Comments to **Table 3.1**	Models can be configured with a variety of peripheral devices, system configurations, and software.
Dispensing volumes	100 nL to 80 μL.
Method of dispensing	Reservoir solutions: syringe-based contact dispensing. Protein: solenoid-based, noncontact dispensing.
Examples of reported %CVs	Typical CVs for 25% PEG 8000 are 5% at 1 μL and 6% at 500 nL.
Number of plates that can be loaded "walk-away"	5 plates: Screenmaker™ 96 +8. 2 plates: Nanodrop™ II. 1 plate: Nanodrop™ ExtY.
Non-SBS formats, 24- or 48-well plates accepted?	Independent dispense allows access to 24- or 48-well plates.
Humidity control?	Not required for Screenmaker™ 96 +8; performance in 42 s. Off-the-shelf humidity chambers or cover-slide system options for the Nanodrop™ .
Chilling available?	Chilling can be achieved with Peltier or recirculating chillers.
Speed	After the robot fills the reservoirs, dispense times for the protein/mother liquor droplets are: 42 s: Screenmaker™ 96 +8; <10 min: Nanodrop™ II.
Associated costs, consumables required, breakable items, etc.	Tip harnesses/tip assemblies. No consumables required except helium and deionized water.
Size	102 × 107 × 63.5 cm; 91 kg.

Information and picture generously provided by Jim Johnson, Innovadyne Technologies, Inc., USA.

Manufacturer: Matrix Technologies Corp., Ltd. **Robot:** Hydra® II Plus-One System

Description	The Hydra® II consists of an array head with 96 syringes for simultaneous delivery of the mother liquors. It is integrated with a single-channel dispenser, the Plus-One, that delivers the protein.
Comments to **Table 3.1**	The Hydra® II is a complete redesign of the original Hydra, which was available for 10 years. The Hydra® II's principle advantage is variable speed control. This means viscous solutions and low volumes can be pipetted much slower, while washing steps, for instance, can be pipetted twice as fast as in the original Hydra. The new unit is also less expensive to build (and buy).
Dispensing volumes	The Plus-One can aspirate 240 μL and is calibrated to dispense down to 50 nL. The 96-head Hydra® II aspirates up to 100 μL and dispenses down to 100 nL. It can be supplied with larger syringes to cover other volume ranges.
Method of dispensing	Hydra® II: positive displacement. The syringes are available with either Teflon-coated stainless-steel needles or Duraflex (a flexible, titanium-based alloy). Plus-One: noncontact dispensing (supplied by Innovadyne).
Examples of reported %CVs	CV of 10% for 200 nL dispenses of 30% PEG 8000.
Number of plates that can be loaded "walk-away"	One. Can be integrated with a plate stacker.
Non-SBS formats, 24- or 48-well plates accepted?	No.
Humidity control?	No. Not necessary because of the speed.
Chilling available?	Yes. Up to four protein samples can be chilled.
Speed	66 s for a 96-well plate (56 s for the mother liquor droplets and 10 s to add the protein). Filling the reservoirs from bulk containers: typically 30 s.
Associated costs, consumables required, breakable items, etc.	Replacement of Teflon seals in the syringes (US$400). Glass syringes can break if a plate is put in upside down, for example. Requires helium.
Size	Hydra® II is 25 × 30 cm; 15 kg. Combined with the Plus-One module, it is approximately 40 × 40 cm.

Information and picture generously provided by Paul Upton, Matrix Technologies Corp., Ltd., USA, and Techtum AB, (their Swedish distributor).

Manufacturer: RoboDesign, Inc.	**Robot:** Robot: Alchemist I™
Description	The Alchemist I™ is a fully automated system for cocktail production/screen formulation, which utilizes RoboDesign's patented "Bird-Feeder™" technology. This novel approach does not require tubing, pumps, or system liquid thus eliminating cross-contamination and increasing reliability. The Alchemist I™ can be purchased as a stand-alone robot or as part of the RoboDesign's complete platform, which also includes: Kalypso™ (for dispensing the protein and mother-liquor droplets and filling the reservoirs), CrystalTrak™ (crystallization database software for optimization and experiment design), and CrystalMation™ (automated visual recognition). Technical specifications for the Kalypso™ are currently being written by the manufacturers.
Comments to **Table 3.1**	Custom crystallization screens can be formulated into either Deep Well blocks or SBS-plates. Each chemical is stored and dispensed via its own individual Bird-Feeder™ and positive displacement pipette combination. This innovative design ensures that no part of the system ever comes in contact with other chemicals. Instead of a multitude of pumps, each Bird-Feeder™ is identified by a unique barcode. Up to 72 different chemicals can be used by the Alchemist I at any one time.
Dispensing volumes	1 mL to 10 mL.
Method of dispensing	Positive displacement syringes.
Examples of reported %CVs	<5% overall CVs for 1 to 10 mL. The use of positive displace ment pipettes allows for verifiable and highly accurate dispensing of a wide range of chemicals. Anything from alcohol to water to 50% PEG 8000 can be dispensed reliably.
Number of plates that can be loaded "walk-away"	The stand-alone version of the Alchemist™ can accommodate one plate at a time. If the Alchemist™ is purchased as part of CrystalMation™, a sufficient number of plates can be handled to allow for walk-away operation for up to 8 h continuously.
Non-SBS formats, 24- or 48-well plates accepted?	Yes.
Humidity control?	No.
Chilling available?	No.
Speed	Formulation and dispensing, e.g., into a Deep Well block, of 96 × 1 ml solutions in under 30 min.
Associated costs, consumables required, breakable items, etc.	"Bird-feeders" may be re-used or even washed but, if required, additional "bird-feeders" can be purchased for approximately US$16 each.
Size	76 x 48 x 76 cm (benchtop model); 152 × 122 × 76 cm (stand-alone); 227 kg.

Information and picture generously provided by Stacy Braunston, RoboDesign International, Inc., USA.

Manufacturer: Tecan Group, Ltd. **Robot series:** Freedom EVO®

Description	A versatile workstation built up of different modules. Because of its modular construction, functions such as microplate shakers, sealers or plate carriers, etc. can easily be incorporated onto the platform as the need arises. It can work from a wide range of tubes, vials, and flasks to generate cocktails. Large volume and submicroliter pipetting can all be handled by the same instrument platform.
Comments to **Table 3.1**	The tips have independent height control and flexible spacing which allows them to accommodate many different bulk containers e.g., 15- or 50-mL Falcon tubes, 50- or 100-mL Pyrex flasks, HPLC-vials, etc., and the different liquid levels in them. The robot can be fitted with a module, the Te-Flipper, for full automation of hanging drops on coverslips. It uses four vacuum plungers, which spot, move, and position the coverslips onto pregreased 48-well plates.
Dispensing volumes	100 nL to 2.5 mL.
Method of dispensing	Contact. Syringes (4 or 8, depending on the model) with Teflon-coated, stainless-steel tips are filled with degassed water that acts as the piston. An air gap of a few microliters separates the col umn of water from the solution to be pipetted. After each dis pense, water is pumped through the syringes to flush and clean them. New water is pumped in to regenerate the liquid piston.
Examples of reported %CVs	CV of 10% for 100 nL of 40% PEG 10,000 (dry transfer).
Number of plates that can be loaded "walk-away"	Unlimited, if sealer and storage modules are included.
Non-SBS formats, 24- or 48-well plates accepted?	Carriers and adapters are available for non-SBS formats (Terasaki plate, VDX plate). The number of wells can be defined in the software.
Humidity control?	Yes, a sliding polystyrene cover is provided as a standard fitting for the carrier that holds the sitting-drop plates. It automatically covers each completed column on the 96-well plate so that the longest time any protein/mother liquor droplet is vulnerable to evaporation is 30 s. Deep Well blocks can also be fitted with an evaporation cover (optional).
Chilling available?	Chilling carriers can be attached to three of the different modules, those for the mother liquors, the protein, and the Deep Well blocks.
Speed	50 min per 96/3-well plate; this includes filling reservoirs.
Associated costs, consumables required, breakable items, etc.	The robot uses fixed steel tips and there are no costs for disposable tips. Deionized water purged with helium is used as system liquid.
Size	Workstations of 1, 1.5, and 2 m length upon which the different modules are assembled.

Information and picture generously provided by Dr. Roland Durner, Tecan Group, Ltd., Männedorf, Switzerland.

Manufacturer: TTP Lab Tech, Ltd.	**Robot:** Mosquito™
Description	The Mosquito™ is a compact nanoliter-pipetting robot utiliz ing a reel of disposable, positive-displacement pipettes.
Comments to **Table 3.1**	Hanging, sitting, and microbatch drops; reservoirs must be prefilled. To make hanging drops, the protein and reservoir solutions are spotted onto a self-adhesive, clear PVC or polycarbonate film, available from the distributor. Ordinary 96-well microtiter plates hold the bulk reservoir solutions (100–300 μL). The user must flip the film and position it onto the microtiter plate with the alignment jig provided.
Dispensing volumes	50 nL to 1.2 μL.
Method of dispensing	Contact dispensing by positive displacement. Single-use, stainless-steel pistons sheathed in high-density polyethylene barrels are mounted on a reel which feeds them individually forward. The moving piston draws up the liquid, then dispenses it by touch-off at the destination plate. The piston pins are made of an inert, low-carbon stainless steel.
Examples of reported %CVs.	Overall CV (dry transfer) for volumes of 100–400 nL is <8%. CV of 6% for 30% PEG 4000. CV of 8% for 10% PEG 20000.
Number of plates that can be loaded "walk-away"	Platform options for 2 or 5 plates. Can be interfaced to a plate-stacking robot.
Non-SBS formats, 24- or 48-well plates accepted?	No.
Humidity control?	Not necessary, due to the speed; a cover is available to provide passive humidification.
Chilling available?	No. However, the hanging-drop blocks and the strip holder for the protein can be pre-chilled.
Speed	Dispensing the reservoir and protein solutions requires: <2 min for 96-wells. <3 min for 384-wells. Tips are disposable; no washing time necessary.
Associated costs, consumables required, breakable items, etc.	Reels of replacement tips. A reel contains 26,000 tips (9-mm pitch), sufficient for 250 96-well plates or 36,000 (4.5-mm pitch) sufficient for 90 384-well plates. Reel replacement on the bandolier is accomplishable by the user in a few minutes. Requires no other equipment, priming solu tions, etc., except two electric outlets.
Size	40 × 65 × 50 cm; 27 kg.

Information and picture generously provided by the distributor for Mosquito, Tony Savill, Molecular Dimensions, Ltd., United Kingdom.

4

Soft-Polymer Microfluidic Applications to Protein Crystallization

James M. Berger

Abstract

The ability to determine structures of biological macromolecules using X-ray crystallography is fundamentally limited by the availability of diffraction quality crystals. Recent developments using microfluidic devices provide a new solution to this problem. Modern day soft-polymer microfluidic designs offer high levels of fluid routing and combining capabilities that can be integrated into small, chip-based devices. Current platforms not only accommodate favored crystal screening techniques, including microbatch, vapor diffusion, and free-interface diffusion, but dramatically lower the amount of material needed for structural studies while accelerating successful screening and imaging efforts. Horizon microfluidic methods, which can interrogate protein phase space, optimize crystal quality, and collect diffraction data *in situ*, stand to dramatically expand the ease and speed with which researchers can identify and scale up crystallization conditions as a prelude to structure determination efforts.

4

Soft-Polymer Microfluidic Applications to Protein Crystallization

James M. Berger

University of California at Berkeley, Berkeley, California

4.1. Introduction

X-ray crystallography has played a central role in dissecting and understanding the molecular function of biological macromolecules. Indeed, crystallographic studies are today arguably the method of choice for determining vast majority of protein and nucleic acid structures, a track record that has arisen from several factors. One is that crystallography permits a broad size range of targets (kDa–MDa) to be imaged from moderate to atomic level resolution. Another is that the molecular models provided by crystallography are biologically relevant, since crystals maintain proteins and nucleic acids in a hydrated state that preserves native shape and catalytic function.

Despite its widespread use, crystallography suffers from some significant technical limitations. The most significant of these is the need for crystals. Because it is currently impossible to predict crystallization conditions for a given target *a priori*, crystal production is rapidly becoming a major bottleneck, particularly so as the demand for structures continues to increase. To complicate matters further, the discovery of crystallization conditions is typically only the first step in the journey toward producing a structure. More often than not, significant effort must also be expended to suitably optimize crystal size, quality, and harvesting conditions for diffraction studies.

Crystallization success depends on the ability to coax a soluble target macromolecule into an ordered state by gently manipulating chemical and environmental variables that influence the target's solution behavior. Crystallization trials are usually conducted by mixing a concentrated solution containing the target macromolecule with solutions of precipitating agents, buffers, and salts. The set of conditions that may produce crystals are empirically determined and must be gleaned from a virtually inexhaustible variety of chemistries. Semirational approaches, such "grid-based," "sparse-matrix," "incomplete factorial," "precipitant synergy," "probabilistic," or "clear strategy" chemical screens are usually applied to rapidly sample crystallization variables, and aim to reduce the complexity of search.[1-7] Numerous equilibration methods for growing crystals of biological macromolecules and complexes also have been developed to streamline and simplify screening, the details of which are reviewed extensively elsewhere[4,8] (see also relevant chapters in this book).

Remarkably, global approaches to crystal screening have changed relatively little over the past several decades. Although setting up several dozen crystallization trials at a time is relatively fast and straightforward, many structure projects require considerably more effort. It is not uncommon for single researcher to spend months or years putting multiple homologs, protein–protein or protein–ligand complexes, truncation variants, and mutants of a single protein through screens in the hopes of finding a single suitable crystal. Even if only 500–1000 conditions are screened for a potential target, perhaps also repeating the screens at a couple of different temperatures, the process rapidly becomes burdensome. In addition, when working with rare or difficult targets such as membrane proteins or protein complexes, the need to obtain 5–10 mg of each target for conducting a thorough screen by traditional manual approaches can be prohibitively laborious.

Robotic systems that work on submicroliter-volume scales have been developed to circumvent some of these issues. These platforms, however, also have their downside. The most sophisticated and high-throughput systems are expensive, and can require significant amounts of space, maintenance, and even dedicated technical staff. Perhaps more importantly, the enhanced speed and reduced sample requirements afforded by robotic crystallization platforms have not fundamentally altered the sampling, setup, or equilibration methods used to grow crystals[9-27] (see also pertinent chapters in this book).

Over the past decade, breakthroughs in materials sciences and engineering have come together to offer an exciting and viable new direction for growing crystals. Termed microfluidics, this technology allows for hundreds to thousands of fluid-routing and mixing events to be conducted combinatorially within a single integrated device using only microliters of sample. Microfluidic devices are ideal for performing liquid handling tasks in parallel and can generate enormous savings in labor, setup time, and space. With respect to crystallization, microfluidics have been used with great success to streamline screening and growth strategies while minimizing sample use. Some of these devices have incorporated design elements that allow direct and unparalleled control of solution mixing and equilibration, while others have begun to make inroads toward overcoming hurdles associated with crystal optimization, harvesting, and diffraction studies.

This review will cover some of the most recent developments in soft-polymer microfluidics and their application to crystallization. In particular, it will attempt to compare and contrast microfluidic approaches with more classic methods, and will provide an overview of the exciting promise and future of this rapidly evolving field.

4.2. Classic Crystal Growth Methods

The search for crystals typically begins with a highly purified sample that has been concentrated to a soluble but supersaturated state. Precipitants are then used to drive the target macromolecule out of solution in a controlled manner that is commensurate with crystal formation. Crystal formation requires precise thermodynamic conditions to generate ordered microcrystalline seeds (nuclei) that are then expanded by the deposition of new molecules onto the growing surface of the crystal lattice. Unfortunately, these conditions are not readily met by most sets of crystallization trial variables, and the most frequent outcome from screening trials is the precipitation of the sample into a randomly aggregated state that is useless for high-resolution diffraction experiments. As a consequence, successful nucleation and subsequent crystal growth depends on careful and systematic exploration of solubility phase space of the target macromolecule.

Crystallographers use different methods to sample and explore solubility phase space. During the process, trade-offs are made between the thoroughness of the search and the amount of labor expended. Two of the most commonly used methods for crystallizing biological macromolecules rely on vapor diffusion and microbatch equilibration schemes. In vapor-diffusion experiments, a drop containing the target macromolecule is mixed with a drop of a precipitating agent. The mixture is suspended and sealed over a well solution containing the precipitant, allowing the osmotic potentials of the two solutions to equilibrate through the vapor phase. Typically, the concentrations of solutes are lower in the drop than in the well so that the drop dehydrates, thereby increasing its solute concentration to match that of the well and promote crystal nucleation and/or protein precipitation. In the microbatch method, the sample solution and precipitant are mixed together and layered under a water-impermeable or semipermeable oil (for reviews, see[4,12,28-30]). When using impermeable oil, equilibration is immediate, although it may take some time before either nucleation or precipitation occurs. In the case of permeable oil, the drop dehydrates, much like in vapor diffusion but often at a slower rate.

If considered on the phase-space diagram of a protein, one can see that vapor-diffusion and microbatch methods approach crystallization in significantly different ways. A common schematic for representing protein phase space is shown in **Plate 4.1**, which plots the behavior of a hypothetical protein or nucleic acid sample in the presence of a precipitating agent as a function of concentration. There are at least two zones that describe protein solubility behavior. The soluble (S) and precipitation (P) regions of phase space are clearly evident from visual inspection of the crystallization trial. Two other zones, the metastable (M) and labile (L) regions, lie between the S and P areas and sustain crystal nucleation and growth. At lower precipitant concentrations, the target occupies a soluble region of phase space, whereas at higher concentrations of precipitant, it is forced from solution into an insoluble, aggregated state. The concentration of precipitant required to manipulate solubility depends on the target concentration, environmental variables (e.g., temperature, pH, and chemical additives), and target-specific properties such as pI or charge distribution. For the purposes of crystallization, an experimenter typically looks for ratios of sample and precipitant concentrations that perch the target on the edge of the solubility boundary, within the labile or metastable regions.[4,31-34]

A diagram showing the differences by which vapor-diffusion and microbatch methods sample phase space is also shown in **Plate 4.1**. Microbatch screens using impermeable oil require that the crystallization trials land directly within the labile region at the outset. By contrast, vapor-diffusion setups and microbatch trials conducted with semipermeable oils initiate the experiment in a region of protein solubility and then enter the labile/metastable regions as the drop concentrates. In this regard, microbatch experiments, when set under impermeable oil, can be likened to a zero-dimensional search of protein solubility, whereas vapor-diffusion screens allow each screening condition to take a linear, one-dimensional walk through the target's phase space. Despite their limited searching power, both vapor-diffusion and microbatch methods remain the most commonly used manual approaches because they consume relatively little sample (0.5–10 μL per experiment), are amenable to parallel processing in small, multi-well trays, and possess a demonstrated track record of success. Both methods also can be automated, often commensurately with significant savings in sample consumption and labor[9-27] (see also relevant chapters in this book).

A crystallization method that differs from either vapor-diffusion or microbatch techniques is that of free interface diffusion (FID).[35,36] In the original implementation of FID, a small volume of target solution is placed in a microcapillary. Next, a precipitating solution is gently layered onto the target. Diffusion occurs through the interface between the fluids, gradually mixing them together. Initially, a portion of the sample encounters an extremely high concentration of precipitant at the point where the fluids meet, while at the distal end of the crystallization chamber, the target solution encounters no precipitant. As diffusion ensues, the precipitant and target solutions mix over the length of the capillary, creating a continuum of concentration gradients for all solutes (**Plate 4.1**).[4,33,36-41] As a consequence, FID allows a single experiment to sample a broad range of the target's phase space and thus represents an extremely efficient means by which to search for crystallization conditions.

Since Salemme first introduced his microcapillary approach to FID, many variations on this scheme have been implemented. For example, target and precipitant solutions can be separated by a gel plug to slow equilibration and prevent sample dilution as precipitant diffuses into the sample chamber.[42-44] Free interface diffusion has also been used extensively in microgravity environments.[45-50] Unfortunately, despite the efficiency of the method for screening phase space, FID has not been as widely adopted by the crystallographic community as either microbatch or vapor-diffusion methods. One reason is that FID setups typically consume relatively large amounts of sample (1–10 μL) to screen a single crystallization condition. Another is that it is difficult to set up, visualize, and store extensive numbers of capillary-based crystallization experiments.

4.3. Microfluidic Platforms for Crystallization

In the past few years, microfluidic technologies have been used to circumvent the limitations inherent to all of these classic approaches. Microfluidic platforms take on many guises, but most of them attempt to conduct combinatorial sampling of multiple solution conditions within a single integrated device using nanoliter-sample volumes. Early microfluidic devices relied on hard materials such as silicon and typically contained only a relatively small number of fluid lines and/or mixing chambers, in part because of the need to build in hard-wired outlet valves or flushing lines to allow fluids and gases to be vented. Despite this added complexity, hard silicon fluidic devices nonetheless have been successfully applied to protein crystallization.[51,52]

More recently, researchers have been making use devices constructed from soft silicone elastomers such polydimethylsiloxane (PDMS). These materials have the advantage that they are biocompatible, transparent, and easy to fabricate into monolithic devices through standard soft-layer lithography techniques.[53-55] The use of PDMS also enables the design of fluid chambers and plumbing lines that may be dead-end loaded by pushing occupying atmosphere into a gas-permeable matrix under pressure. This property obviates the need for on-board venting valves and greatly simplifies microfluidic chip design, allowing users to implement much higher line densities and complex fluid routing schemes than can be achieved by hard-material approaches.[53,56]

As a consequence, the processing power of microfluidics technologies has become quite formidable. Currently, thousands to tens of thousands of individual fluid lines, valves, and chambers can be integrated and individually actuated or interrogated on a single device only a few centimeters in size.[57,58] Given the empirical nature of crystallization, together with the sample limitations associated with biological macromolecules, the parallel sampling power of microfluidics is perfectly suited to conducting screening trials. Just as multiple methods have evolved in the laboratory for crystallizing proteins using standard approaches, multiple soft-polymer microfluidic approaches to grappling with the problem of protein crystallization and phase-space screening also have been developed.

One scheme, recently described by Ismagilov and colleagues, uses microfluidics to mimic microbatch crystallization under oil.[59] In this design, the researchers created a chip that contained fluid lines for four different solutions (**Plate 4.2a**). The fluid lines are arrayed so as to meet and combine at a single junction. By controlling the amount of fluid that is pulsed by each line into the merging chamber, the various ratios of sample, precipitant, and buffer can be systematically altered. The composite solution is then routed into a new fluid line that contains a water-immiscible fluorinated oil[59] to promote the formation of reverse micelles and permit solute mixing by chaotic advection.[60,61] Once set up, the droplets are allowed to equilibrate in the chip where they can be visualized to assess crystal growth (**Plate 4.2b**).

It is interesting to note that the microbatch chip also serves as a type of liquid-handling station, capable of melding a large number of mixing ratios into a grid-based screen that can provide thorough coverage of a sample's response to a single precipitant. More recently, the Ismagilov group described a variation on this device that allows for the setup of "vapor-diffusion-like" experiments. The chip uses a small plug of water-permeable oil to separate a drop containing protein and precipitant from another solution containing a high concentration of salt (**Plate 4.2c**).[62] Water migrates from the target–precipitant drop into the high-salt solution, thereby dehydrating the drop and increasing the solute concentrations to promote crystal nucleation and growth.

Another PDMS-based microfluidic device, reported by Quake and colleagues, takes a significantly different approach for crystallization by utilizing FID.[56] The platform permits the screening of 48 distinct precipitant solutions against a single protein sample and manifests three different mixing ratios of protein to each precipitant to increase coverage of solubility phase space (**Plate 4.3a**). Trials on this microfluidic chip work by isolating the protein and precipitant solutions in separate, but paired chambers (**Plate 4.3b**). The chambers are connected by a thin channel that is kept closed during loading by the action of an interface valve located midway between the two chambers. Following loading, the chambers are sealed off by containment valves, and the interface valve between them is opened to allow free interface diffusion to occur.

The robustness and capabilities of the FID chip for crystallization are evident from a number of standpoints. For one, the device requires very little sample, using less than 3 μL of material to explore 144 different reagent conditions (48 conditions × 3 concentration ratios). Indeed, over the course of six days, a single researcher was able screen thousands of crystallization conditions for a dozen targets while consuming less than a hundred microliters of total sample (**Plate 4.3c**).[56] Examples shown in the figure include: (*i—Left*) *Aquifex aeolicus* DnaA—these crystals are only one-third smaller than those grown by vapor-diffusion methods for structure determination,[76] even though the volumes of solutions used in the two instances differ by 40 fold (25 nL in chip vs. 1 μL by hanging drop); (*ii—Upper right*) the B-subunit of archaeal topoisomerase VI obtained from sparse matrix screens and grown in a 12 nL chamber;[56,77] and (*iii—Lower right*) RNase III from *M. tuberculosis* (inset shows 2× higher magnification). In the case of *Mtb* RNAseIII, this crystal form appeared in only one of the three mixing ratios from a sparse matrix screen. Prior screens using traditional methods failed to identify this condition, whereas follow-up experiments based on the chip conditions not only successfully reproduced the hit but also revealed that crystal growth occurred only in a narrow range of PEG/$CaCl_2$/protein ratios. This was the only hit obtained for this protein from over 600 different screening conditions.[56,79]

In addition to low sample consumption, the FID chip gently introduces the target to precipitant. This property eliminates the "solubility shock" experienced by macromolecules encounter dramatic changes in their solution environment and reduces inadvertent precipitation. Perhaps most importantly, however, FID chips were seen to produce crystals for an equal or greater number of conditions as compared to screens set by traditional methods. The setup even supported the growth of large, single, diffracting crystals from nanoliter volumes of target solution.[56] For example, 100-μm crystals of thaumatin were grown from as little as 5 nL of solution. When extracted from the chip and frozen, these crystals diffracted to 1.3 Å.

Second-generation FID chips have been produced and marketed as part of the TOPAZ™ system by Fluidigm Corporation (www.fluidigm.com). These devices, collectively termed "X.96" chips (where "X" represents the number of independent targets that can be accommodated by a single chip), allow FID to take place in a single 0.75-nL sample channel connected to a 2.6-nL crystallization reagent well (**Plate 4.4**). Extensive valving regulated by an automated loading station actuates control lines to both load the chip and initiate solution mixing without user intervention. The X.96 chip design allows for extremely high densities of experiments to be contained within and screened by a single device. Current TOPAZ™ chips can accommodate either 96 ("1.96") or 384 ("4.96") independent experiments in a single 2 × 5 cm block of PDMS, and they consume only 1.5–6 μL of sample (**Plate 4.4**). The TOPAZ™ device also has a small footprint (12 × 8.5 × 1.5 cm) and is compatible with industry-standard robotic pipetting devices for automating chip setup.

4.4. Capabilities Unique to Microfluidic Devices

One major advantage of microfluidics over manual or automated pipetting systems lies in its power to rapidly and accurately dispense highly viscous solutions and very small sample volumes. A second resides in its ability to combinatorially mix together large numbers of solutions with little user time or effort. For example, the microbatch and vapor-diffusion crystallization chips constructed by Zheng et al. require only four input solutions but permit the ratios of protein to precipitant to be as finely sampled as the user desires.[59,62] Such increased coverage means that a given sample can be interrogated more robustly, which represents a dramatic improvement over manual approaches for crystal growth and optimization. Crystallization chips based on FID offer similar fluid dispensing and sample consumption advantages. Although current FID chips do not yet act as minimized reagent mixing stations, they do allow a continuum of target to precipitant ratios to be sampled for a defined set of starting solutions, thereby increasing the chances of encountering the critical regions of phase space required for crystal growth.[4,33,36-41,56] Indeed, FID chips have already been used to successfully crystallize dozens of different non-model targets, ranging from 12 kDa to 2 MDa in size.[56,63] * In one recent and highly significant study, Segelke has found that microfluidic FID chips had approximately 2–10-fold higher success rate than an automated vapor-diffusion robot for identifying crystallization conditions for five novel protein targets.[80]

PDMS-based μFID devices also can take advantage of the water-permeable properties of the polymer to respond to the relative humidity of their environment and route crystallization screens through additional regions of solubility phase space. For example, following solution mixing by FID, a PDMS chip can be incubated at high humidity to equilibrate into a uniform single concentration of protein and precipitant and maintain that state. Alternatively, the chip can be modestly dehydrated to further concentrate solutes in the reaction chambers. Fluidigm X.96 chips incorporate a hydration chamber that surrounds the PDMS, allowing for continuous control over the hydration state of all experiments on the chip.

Consider, for example, the scheme shown in **Plate 4.5a**. This graph shows the relative diffusion times of different molecular-weight components across a 400 μm long channel of 100 × 10 μm cross section that connects two 400 × 400 × 100 μm chambers. Salts are of low molecular weight and, therefore, diffuse rapidly throughout the chambers to reach a steady state concentration in a matter of a few hours. PEG 3350, which is of larger mass, reaches nearly full equilibration in under a day. By contrast, less than 20% of the concentration of a 40 kDa protein changes in the sample chambers even after two days.

By opening the interface valve separating the sample and precipitant chambers for only a fixed amount of time, on the order of around 12 h for this model setup, a target therefore can be exposed to nearly the full concentration of precipitant without becoming significantly diluted itself. **Plate 4.5b** illustrates how FID chips can efficiently hone in on appropriate conditions of phase space for a sample with a small metastable/labile (M/L) zone (such as we observed for *M. tuberculosis* RNase III, **Plate 4.3c**) without sacrificing the high concentrations of sample needed to maintain supersaturation. If the entire length of the

* A. May, personal communication

FID channel is taken into consideration for a given period of time (typically hours), then the experiment samples a two-dimensional continuum of phase space (thick blue lines bounding a light blue area). The total area covered by the curves is only qualitatively illustrated—real curves are much more complex and depend on a host of variables including channel dimensions, the duration of mixing, and solute diffusion rates. Moreover, the interface line between reagent and sample chambers can be closed before equilibrium has been attained, keeping the sample concentration high, shifting the final target concentration value upward from its midpoint on one side of the channel and leaving a lower sample concentration on the other (*middle blue circles*). The region between the two endstates is transiently sampled during equilibration (*dashed line*). If the M/L zone resides near any of these sample concentrations (i.e., supersaturation values), this scheme allows the experiment to pass through it.

If, following valve closure, the chip is placed in a modestly dehydrating environment, the sample and precipitant will concentrate together further. This property gives the experiment yet another chance to pass into regions of phase space required for crystal nucleation and growth (**Plate 4.5c**). Thus, FID chips allow each crystallization condition to interrogate two distinct regions of phase space and yet maintain high macromolecule concentrations. Such precise and reproducible kinetic control is extremely difficult to match using traditional methods and is virtually impossible to achieve on any type of large or high-throughput experimental scale.

Another advantage of microfluidic chips over traditional formats is that the small length scales accessed by these devices are ideally suited to counteract the buoyant and convective forces that are often present in the larger volume setups.[33,38,47,49,56,64] For FID, this property means that a stable free-fluid interface may be easily established and that crystals will tend to remain in a fixed position, within their ideal growth zone, as their mass increases. Movement of solutes in and around the crystals also will be subject principally to diffusive and not to convective forces. Previously, the only way in which these situations could be attained was to set up a FID experiment in microgravity environments or by using gels.[33,38,46-50,64-66] This behavior holds the prospect that it may be possible to grow crystals in microfluidic devices that are of extremely high quality,[81] as has been reported for those produced under microgravity environments.

4.5. Microfluidic Phase-Space Mapping

The microfluidic systems described thus far provide a new, highly convenient format with which to explore crystallization conditions more systematically and with less sample material than is easily attainable by traditional approaches. However, none of these platforms actually change the empirical search methods used to identify crystallization conditions. Recently, Quake and colleagues have employed microfluidics with great success toward overcoming this problem.[67] Their new type of chip, termed a "formulator," relies on a rotary pump to systematically and combinatorially mix dozens of different precipitants and buffers with a sample solution. The system reports back on whether a set of conditions precipitates the sample, providing a complete map of the solubility boundaries of the target for a given set of reagents in a "one-shot" experiment (**Plates 4.6a** and **b**). Because the metastable/labile regions desired for crystal nucleation and growth lie at the interface of these precipitation boundaries, the formulator can identify the optimal types and concentrations of both precipitant and buffer to be used in crystallization screening.

Hansen et al. used the formulator to successfully map out the boundaries between the soluble and precipitated regions of phase space for a set of test proteins and a bank of 4,300 reagent conditions (**Plates 4.6c** and **d**).[67] From this information, the authors then derived a highly focused crystallization screen for xylanase that increased the efficiency of their search for crystallization conditions by approximately 80 fold over random screening methods. Remarkably, data from the formulator yielded new conditions that had not been described previously for the enzyme. Innovations such as the formulator, as well as other nanodroplet screening methods (e.g., acoustic levitation and piezoelectric mixing[68]) are thus moving real-time phase-space mapping into the realm of benchtop feasibility, and they promise to change the "hit-or-miss" process of crystallization into a more rational endeavor.

4.6. Horizon Methods

In many respects, the field of microfluidics and its application to structural biology is just in its infancy. Since one of the most significant capabilities of microfluidics lies in its ability to create and work with scaleable architectures, one can expect over the long run that the number of crystallization experiments conducted on board a single, "one-shot" device will continue to increase, without commensurate consumption of sample. Such high-density arrays have already been implemented as a formal demonstration that chips can serve as addressable memory devices.[57] With respect to crystallization, new 8.96 chips developed by Fluidigm Corp. can embed nearly 800 distinct experiments in a single chip. Combinatorial solution mixers such as the formulator promise similar scales of improvement in the efficiency of liquid handling, reagent mixing, and protein phase-space mapping.

The malleable design of microfluidic devices also opens new windows into screening crystallization parameters that are normally difficult access by traditional methods. For example, pH often influences crystallization propensity and can be used as a primary driving force for the growth of protein crystals. [69,70] Unfortunately, the control of pH is somewhat cumbersome to achieve. Traditionally, pH is changed in a crystallization setup either through slow dialysis or by the use of volatile salts such as ammonium acetate in a vapor-diffusion experiment.[4] Microfluidics, through the ability to engineer a variety of connection channel lengths and diameters between sample and reagent chambers, offers a new means to more efficiently and systematically explore pH as a crystallization approach.

Temperature is a similarly under-utilized parameter. Most researchers conduct crystallization trials at standard temperatures such as 20° and 4°C. Occasionally, incubators are used to target a more specific temperature; however, researchers are then faced with the problem of examining crystals at those temperatures. In addition, if one wants to experiment with slow temperature changes and gradients as a driving force to effect nucleation or crystal growth, an entire 24- or 96-well tray must generally be subjected to the process.[69,71] By contrast, microfluidic chips can be used and visualized on flat temperature thermocyclers, and customized chips even can be constructed with embedded temperature controllers for finer resolution with position-specific control.[72-75] In one recent example applied to crystallization, Cooper and co-workers used a medium volume hard silicon chip to study lysozyme crystallization under the control of defined temperature gradients across a series of microbatch wells.[51] Future chip designs should allow temperature gradients and ramping cycles to be queried more thoroughly and to be applied individually to a number of conditions in a manner that does not require incubators.

Finally, significant strides are being made toward overcoming bottlenecks associated with crystal optimization and diffraction screening. For example, two devices have been produced that allow for X-ray diffraction to be assessed from crystals *in situ*. In one design, Zheng et al. interfaced their vapor-diffusion microfluidic chip with a glass capillary to grow and place unfrozen crystals in an X-ray beam without the need for harvesting (**Plates 4.7a–b**).[62] With this device, the researchers demonstrated diffraction directly from their microfluidics chamber on a crystal of thaumatin to 1.8 Å resolution. Using a different approach, both the Quake laboratory and Fluidigm Corp. have developed prototype thin-layer PDMS disks (< 150 μm thick) that permit single crystallization chambers to be placed into an X-ray beam (**Plates 4.7c–e**). *In situ* diffraction has been obtained from a variety of unfrozen crystals grown in these devices to resolutions as high as 1.2 Å (lysozyme).[81,*] Notably, crystals within the disks can be soaked with cryoprotectant and frozen directly in liquid nitrogen to extend crystal lifetime in the beam.

It is important to note that these diffraction chip designs are compatible with the larger sample volumes (around 100–200 nL) typically needed to produce crystals of sufficient size for data collection (**Plate 4.7d**). The FID-based diffraction chips also have incorporated sophisticated interface-line arrangements between sample and reagent chambers to alter the rate of fluid mixing and tightly control equilibration kinetics. For example, **Plate 4.7e** diagrams a prototype TOPAZ™ "X-ray" chip. The sample chamber (narrow rectangle with blue/white gradient) is connected to the reagent chamber (large blue rectangle) by a variable number of interface lines (1–4 lines per condition, sectors connected by two or three lines are shown). These wells are embedded in a PDMS chip that mounts into the same carrier shown in **Plate 4.4a**. The maximal number of connecting interface lines (four) recreates the exact equilibration kinetics as attained in the 1.96 and 4.96 FID chambers ($\tau = 5/5$), whereas reducing the number of interface lines to three, two, or one slows down equilibration by 75%, 45%, and 20% ($\tau = 5/4$, 5/2, and 5/1). Crystal size, number, and quality can all be seen to vary as a function of equilibration time regulated by different numbers of interface lines (**Plate 4.7f**).

---

* K. Franke, H., Nassef, J. Holton, *unpublished*

4.7. Conclusion

Overall, through the targeted development of robust screening, optimization, harvesting, and diffraction regimes, microfluidics are reaching the point of becoming a unified approach that is capable of moving all the way from purified sample to structure. Although there are some limitations surrounding microfluidic approaches to protein crystallization, most notably with respect to cost and design flexibility, it is important to note that the field of microfluidics has only been vigorously applied to the crystallization of biological macromolecules for the past few years. In this respect, microfluidics occupies a deve-lopmental period analogous to that of the early silicon transistor chip. Though expensive and limited in capability at the outset, silicon chips quickly improved to become dramatically more powerful at ever-lower costs. As the technology matures, microfluidics likewise stand to evolve further into a low-cost, off-the-shelf technology that is robust and user friendly, and whose technical benefits (combinatorial mixing, free interface diffusion, etc.) dwarf those of manual or robotic methods. One can anticipate that the same forces that propelled silicon chips to dominate the computer and electronics industries will similarly assist microfluidics becoming the platform of choice for high-throughput protein characterization and crystallization.

Acknowledgements

The author is grateful to C. Hansen, R. Ismagilov, A. May, and S. Quake, as well as Fluidigm, Corp. for providing images for this manuscript and to C. Hansen, K. Frankel, H. Nassef, Brent Segelke, and S. Quake for sharing preliminary data. The author also would like to thank C. Hansen, L. Dodd, A. May, and D. Minor for critical reading of the manuscript and C. Hansen and S. Quake for insightful discussions on FID and microfluidics. This work, as well as some of the proteins and projects featured in it, was supported by grants from the G. Harold and Leila Y. Mathers Charitable Foundation, the David and Lucile Packard Foundation, and the NIH (CA77373 and P50-62410).

References

1. Carter CW Jr. and Carter CW. **Protein crystallization using incomplete factorial experiments.** *J Biol Chem* 1979, **254**(23):12219–12223.

2. Jancarik J and Kim S-H. **Sparse matrix sampling: A screening method for crystallization of proteins.** *J Appl Cryst* 1991, **24**:409–411.

3. Cudney R et al. **Screening and optimization strategies for macromolecular crystal growth.** *Acta Crystallogr D Biol Crystallogr* 1994, **50**:414–423.

4. McPherson A. ***Crystallization of biological macromolecules.*** Cold Spring Harbor, New York: Cold Spring Harbor Laboratory Press, 1999, pp. ix, 586.

5. Hennessy D et al. **Statistical methods for the objective design of screening procedures for macromolecular crystallization.** *Acta Crystallogr D Biol Crystallogr* 2000, **56**(Pt 7):817–827.

6. Brzozowski AM and Walton J. **Clear strategy screens for macromolecular crystallization.** *J Appl Cryst* 2001, **34**:97–101.

7. Majeed S. et al. **Enhancing protein crystallization through precipitant synergy.** *Structure (Camb)* 2003, **11**(9):1061–1070.

8. Ducruix A and Giegé R. ***Crystallization of nucleic acids and proteins: A practical approach.*** 2nd ed. The practical approach series; 210, 1999, Oxford, New York: Oxford University Press, 1999, pp. xxii, 35.

9. Segelke BW et al. **Laboratory scale structural genomics.** *J Struct Funct Genomics* 2004, **5**(1–2):147–157.

10. McPherson A. **Protein crystallization in the structural genomics era.** *J Struct Funct Genomics* 2004, **5**(1–2):3–12.

11. Jacquamet L et al. **A new highly integrated sample environment for protein crystallography.** *Acta Crystallogr D Biol Crystallogr* 2004, **60**(Pt 5):888–894.

12. D'Arcy A, Sweeney AM, and Haber A. **Practical aspects of using the microbatch method in screening conditions for protein crystallization.** *Methods* 2004, **34**(3):323–328.

13. Bard J et al. **Automated systems for protein crystallization.** *Methods* 2004, **34**(3):329–347.

14. Weselak M et al. **Robotics for automated crystal formation and analysis.** *Meth Enzymol* 2003, **368**:45–76.

15. Rupp B. **High-throughput crystallography at an affordable cost: The TB Structural Genomics Consortium Crystallization Facility.** *Acc Chem Res* 2003, **36**(3):173–81.

16. Luft JR et al. **A deliberate approach to screening for initial crystallization conditions of biological macromolecules.** *J Struct Biol* 2003, **142**(1):170–179.

17. Hui R and Edwards A. **High-throughput protein crystallization.** *J Struct Biol,* 2003, **142**(1):154–161.

18. DeLucas LJ et al. **Efficient protein crystallization.** *J Struct Biol* 2003, **142**(1):188–206.

19. Villasenor A et al. **Fast drops: A high-throughput approach for setting up protein crystal screens.** *Biotechniques* 2002, **32**(1):184,186,188–189.

20. Stewart L, Clark R, and Behnke C. **High-throughput crystallization and structure determination in drug discovery.** *Drug Discov Today* 2002, **7**(3):187–196.

21. Kuil ME et al. **Protein nano-crystallogenesis.** *Enzyme Microb Technol,* 2002, **30**(3):262–265.

22. Krupka HI et al. **The high-speed Hydra-Plus-One system for automated high-throughput protein crystallography.** *Acta Crystallogr D Biol Crystallogr* 2002, **58**(Pt 10 Pt 1):1523–1526.

23. Chayen NE and Saridakis E. **Protein crystallization for genomics: Towards high-throughput optimization techniques.** *Acta Crystallogr D Biol Crystallogr* 2002, **58**(Pt 6 Pt 2):921–927.

24. Stevens RC. **High-throughput protein crystallization.** *Curr Opin Struct Biol* 2000, **10**(5):558–563.

25. Mueller U et al. **Development of a technology for automation and miniaturization of protein crystallization.** *J Biotechnol* 2001, **85**(1):7–14.

26. Morris DW, Kim CY, and McPherson A. **Automation of protein crystallization trials: Use of a robot to deliver reagents to a novel multi-chamber vapor diffusion plate.** *Biotechniques* 1989, **7**(5):522–527.

27. Kelders HA et al. **Automated protein crystallization and a new crystal form of a subtilisin:Eglin complex.** *Protein Eng* 1987, **1**(4):301–303.

28. Chayen NE. **Comparative studies of protein crystallization by vapour-diffusion and microbatch techniques.** *Acta Crystallogr D Biol Crystallogr* 1998, **54(Pt 1)**:8–15.

29. Chayen NE, **The role of oil in macromolecular crystallization.** *Structure* 1997, **5**(10):1269–1274.

30. Rayment I. **Small-scale batch crystallization of proteins revisited: An underutilized way to grow large protein crystals.** *Structure (Camb)* 2002, **10**(2):147–151.

31. Ataka M. **Protein crystal growth: An approach based on phase diagram determination.** *Phase Transitions* 1993, **45**:205–219.

32. Blow DM et al. **Control of nucleation of protein crystals.** *Prot Sci* 1994, **3**(10):1638–1643.

33. Garcia-Ruiz JM. **Counterdiffusion methods for macromolecular crystallization.** *Meth Enzymol* 2003, **368**:130–154.

34. Haas C and Drenth J. **Understanding protein crystallization on the basis of the phase diagram.** *J Cryst Growth* 1999, **196**(2–4):388–394.

35. Salemme FR. **Protein crystallization by free interface diffusion.** *Meth Enzymol* 1985, **114**:140–141.

36. Salemme FR. **A free interface diffusion technique for the crystallization of proteins for X-ray crystallography.** *Arch Biochem Biophys* 1972, **151**(2):533–539.

37. Luft JR and DeTitta GT. **Kinetic aspects of macromolecular crystallization.** *Meth Enzymol* 1997, **276**:110–131.

38. Carotenuto L. et al. **Theory and simulation of buoyancy-driven convection around growing protein crystals in microgravity.** *Microgravity Sci Technol* 2002, **13**(3):14–21.

39. Carotenuto L et al. **Experimental observations and numerical modelling of diffusion-driven crystallisation processes.** *Acta Crystallogr D Biol Crystallogr* 2002, **58**(Pt 10 Pt 1):1628–1632.

40. Garcia-Ruiz JM, Novella ML, and Otalora F. **Supersaturation patterns in counter-diffusion crystallisation methods followed by Mach-Zehnder interferometry.** *J Cryst Growth* 1999, **196**(2–4):703–710.

41. Cang HX and Bi RC. **Influence of various factors on the liquid/liquid diffusion crystallization of proteins.** *J Cryst Growth* 1999, **196**(2–4):442–446.

42. Moreno A and Soriano-Garcia M. **Crystal-growth kinetics of protein single crystals along capillary tubes in the gel-acupuncture technique.** *Acta Crystallogr D Biol Crystallogr* 1999, **55 (Pt 2)**:577–580.

43. Garcia-Ruiz JM. **Uses of crystal growth in gels and other diffusing-reacting systems.** *Key Eng Mater* 1991, **88**:87–106.

44. Garcia-Ruiz JM. **Investigations on protein crystal growth by the gel acupuncture method.** *Acta Crystallogr D Biol Crystallogr* 1994, **50**(Pt 4):484–490.

45. Chayen NE and Helliwell JR. **Microgravity protein crystallization: Are we reaping the full benefit of outer space?** *Ann NY Acad Sci* 2002, **974**:591–597.

46. DeLucas LJ, Moore KM, and Long MM. **Protein crystal growth and the International Space Station.** *Gravit Space Biol Bull* 1999, **12**(2):39–45.

47. DeLucas LJ et al. **Protein crystal growth in microgravity.** *Science* 1989, **246**(4930):651–654.

48. Garcia-Ruiz JM et al. **Granada Crystallisation Box: A new device for protein crystallisation by counter-diffusion techniques.** *Acta Crystallogr D Biol Crystallogr* 2002, **58**(Pt 10 Pt 1):1638–1642.

49. Wang YP et al. **Protein crystal growth in microgravity using a liquid/liquid diffusion method.** *Microgravity Sci Technol* 1996, **9**(4):281–283.

50. Otalora F et al. **Growth of lysozyme crystals under microgravity conditions in the LMS (STS-78) mission.** *J Cryst Growth* 1999, **196**(2-4):649–664.

51. Juarez-Martinez G et al. **High-throughput screens for postgenomics: Studies of protein crystallization using microsystems technology.** *Anal Chem* 2002, **74**(14):3505–3510.

52. Sanjoh A and Tsukihara T. **Spatiotemporal protein crystal growth studies using microfluidic silicon devices.** *J Cryst Growth* 1999, **196**(2–4):691–702.

53. Unger MA et al. **Monolithic microfabricated valves and pumps by multilayer soft lithography.** *Science* 2000, **288**(5463):113–116.

54. Xia Y et al. **Complex optical surfaces formed by replica molding against elastomeric masters.** *Science* 1996, **273**(5273):347–349.

55. Xia YN and Whitesides GM. **Soft lithography.** *Annual Review of Materials Science* 1998, **28**:153–184.

56. Hansen CL et al. **A robust and scalable microfluidic metering method that allows protein crystal growth by free interface diffusion.** *Proc Natl Acad Sci USA* 2002, **99**(26):16531–16536.

57. Thorsen T, Maerkl SJ, and Quake SR. **Microfluidic large-scale integration.** *Science* 2002, **298**(5593):580–584.

58. Groisman A, Enzelberger M, and Quake SR. **Microfluidic memory and control devices.** *Science* 2003, **300**(5621):955–958.

59. Zheng B, Roach LS, and Ismagilov RF. **Screening of protein crystallization conditions on a microfluidic chip using nanoliter-size droplets.** *J Am Chem Soc* 2003, **125**(37):11170–11171.

60. Thorsen T et al. **Dynamic pattern formation in a vesicle-generating microfluidic device.** *Phys Rev Lett* 2001, **86**(18):4163–4166.

61. Song H and Ismagilov RF. **Millisecond kinetics on a microfluidic chip using nanoliters of reagents.** *J Am Chem Soc* 2003, **125**(47):14613–14619.

62. Zheng B et al. **A droplet-based, composite PDMS/glass capillary microfluidic system for evaluating protein crystallization conditions by microbatch and vapor-diffusion methods with on-chip X-ray diffraction.** *Angew Chem Int Ed Engl* 2004, **43**(19):2508 2511.

63. Hansen C. **Microfluidic technologies for structural biology.** In: ***Applied Physics*** 2004, California Institute of Technology: Pasadena. p. 231.

64. Vekilov PG et al. **Nonlinear dynamics of layer growth and consequences for protein crystal perfection.** *J Cryst Growth* 1999, **196**(2–4):261–275.

65. McPherson A et al. **The effects of microgravity on protein crystallization: Evidence for concentration gradients around growing crystals.** *J Cryst Growth* 1999, **196**(2–4):572–586.

66. Carter DC et al. **Diffusion-controlled crystallization apparatus for microgravity (DCAM): Flight and ground-based applications.** *J Cryst Growth* 1999, **196**(2–4):602–609.

67. Hansen CL, Sommer MOA, and Quake SR. **Systematic investigation of protein phase behavior with a microfluidic formulator.** *Proc Natl Acad Sci USA* 2004, **101**:14431–14436.

68. Santesson S et al. **Screening of nucleation conditions using levitated drops for protein crystallization.** *Anal Chem* 2003, **75**(7):1733–1740.

69. McPherson A **Crystallization of proteins by variation of pH or temperature**. *Meth Enzymol* 1985, **114**:125–127.

70. McPherson A. **Increasing the size of microcrystals by fine sampling of pH limits.** *J Appl Cryst* 1995, **28**:362–365.

71. Luft JR, Rak DM, and DeTitta GT. **Microbatch macromolecular crystallization on a thermal gradient.** *J Cryst Growth* 1999, **196**(2–4):447–449.

72. Liu J, Hansen C, and Quake SR. **Solving the "world-to-chip" interface problem with a microfluidic matrix.** *Anal Chem* 2003, **75**(18):4718–4723.

73. Liu J, Enzelberger M, and Quake S. **A nanoliter rotary device for polymerase chain reaction.** *Electrophoresis* 2002, **23**(10):1531–1536.

74. Mao H, Yang T, and Cremer PS. **A microfluidic device with a linear temperature gradient for parallel and combinatorial measurements.** *J Am Chem Soc* 2002, **124**(16):4432–4435.

75. Mao H, Yang T, and Cremer PS. **Design and characterization of immobilized enzymes in microfluidic systems.** *Anal Chem* 2002, **74**(2):379–385.

76. Erzberger JP, Pirruccello MM, and Berger JM. **The structure of bacterial DnaA: Implications for general mechanisms underlying DNA replication initiation.** *Embo J* 2002. **21**(18): 4763–4773.

77. Corbett KD and Berger JM. **Structure of the topoisomerase VI-B subunit: Implications for type II topoisomerase mechanism and evolution.** *Embo J* 2003, **22**(1):151–163.

78. Skordalakes E and Berger JM. **Structure of the Rho transcription terminator: Mechanism of mRNA recognition and helicase loading.** *Cell* 2003, **114**(1):135–146.

79. Akey D and Berger JM. **Structure of the nuclease domain of ribonuclease III from *M. tuberculosis* at 2.1 Å.** *Prot Sci* 2005, **14**:2744–2750.

80. Segelke B. **Macromolecular crystallization with microfluidic free-interface diffusion.** *Expert Rev Proteomics* 2005, **2**(2):165–172.

81. Hansen CL, Classen SC, Berger JM, and Quake SR. **A micro-fluidic device for kinetic optimization of protein crystallization and *in situ* structure determination**. *J Am Chem Soc* 2006, **128**:3142–3143.

5

Counter-Diffusion Capillary Crystallization for High-Throughput Applications

Juan M. Garcia-Ruiz and Joseph D. Ng

Abstract

Counter-diffusion crystallization in capillaries is an effective means to produce protein crystals suitable for X-ray analysis without harvesting and transporting challenges. We review the principles, general applications, instructions, and recent tools available for protein crystallization screening and optimization using counter diffusion in a restricted geometry. Using this technique, the loading of the protein, equilibration process, addition of strong anomalous scattering atoms, and cryogenic soaking can be simultaneously executed. Crystals obtained by this method can be used for immediate X-ray data analysis and provide quality data sufficient for structure determination without crystal handling. The ease and versatility of the technique has made it a very useful method for high-throughput manipulation.

5

Counter-Diffusion Capillary Crystallization for High-Throughput Applications

Juan M. Garcia-Ruiz[*] and Joseph D. Ng[†]

[*]Concejo Superior de Investigaciones Cientificas and Universidad de Granada, Granada, Spain
[†]University of Alabama in Huntsville, Huntsville, Alabama, USA

5.1. Why Counter-Diffusion Crystallization?

Purified proteins solutions loaded in capillary tubes can be screened for finding optimal crystal growth conditions against a precipitant solution by counter-diffusion method.[1] The counter-diffusion crystallization procedure thereby can be used as an effective means to quickly produce protein crystals suitable for X-ray analysis without harvesting and transporting challenges. Advanced crystallization devices have been constructed to quickly screen or optimize protein crystallization conditions by counter diffusion in a single- or multiple-array configuration. The loading of the protein, equilibration process, addition of strong anomalous scattering atoms, and cryogenic soaking can be simultaneously executed.[2] Crystals obtained by this method can be used for immediate X-ray data collection where incorporation of an anomalous scatterer for *ab initio* structural determination by Single-wavelength Anomalous Scattering (SAS) or by other X-ray crystallographic phasing techniques can be performed without any crystal manipulation.[3] Multiple counter-diffusion processes (such as in 96-tube format) can be performed at one time in an automated manner. These tools and techniques along with the advances in protein expression and purification, X-ray radiation, and crystallographic software packages allow the full streamlining of tedious and time consuming events in determining three-dimensional structures for structural genomics endeavors.

5.2. Principle of Counter-Diffusion Crystallization

Protein crystallization by counter diffusion in a restricted geometry is initiated by preparing a precipitating agent and protein solution to be arranged next to one another inside an X-ray capillary tube. The two solutions are set to diffuse against each other where the protein diffuses slower than the precipitant molecules due to their different molecular weights. Consequently a spatial-temporal gradient of supersaturation along the length of the capillary is formed (**Plate 5.1**). Unlike conventional vapor-diffusion or batch crystallization techniques where just one or a few supersaturation conditions are screened in one experiment, crystals grown by counter-diffusion are obtained from different supersaturation environments created along the capillary.[4] Unlike free-interface-diffusion method, where precipitant and protein solutions are allowed to mix promptly towards equilibrium,[5] the counter-diffusion method initiates precipitation under far equilibrium conditions in one extreme of the capillary while allowing the diffusion-reaction system to slowly self-approach equilibrium along the length of the capillary. Consequently a wide and gradual range of crystallization conditions is formed across the capillary volume in one single experiment.

5.3. Preparation and Setup

The choices of capillary configurations are multiple for counter-diffusion crystallization. A single capillary can be simply used to screen or optimize a crystallization condition using one selected precipitant. Multiple capillaries prepared in tray, cartridge, or 96-capillary formats can be used with sparse-matrix or customized screening solutions containing different reagents and additives.

The protein concentration required for screening of crystallization conditions depends on the overall solubility of the protein and how far the crystallization pH is from its isoelectric point. Successful screenings yielding crystalline precipitate have been performed using concentrations as low as 2 mg/mL.[6] However, a concentration of 5 to 10 mg/mL is recommended when working one or two pH units away from the protein's isoelectric point. For optimization of crystallization conditions, it is recommended to increase the concentration of protein to 20 mg/mL. Note the reason why such low protein concentration is required for screening is because the initial precipitant concentration is recommended to be as high as possible to assure the precipitation of the protein upon contact with the protein solution. The very nature of the method assures the initial presence of an amorphous precipitation would have a tendency to produce crystalline products within the capillary (for review see references[1,3]).

Quartz capillaries (Charles Supper Company, Natick, MA) or borosilicate glass capillaries (Triana Science & Technology, Spain) of equivalent material are most suitable for the counter-diffusion crystallization method. The diameter of the tubes should be no bigger than 0.3 mm without any gel additives to minimize convection and maintenance of a supersaturation gradient during the equilibration process. However larger diameter tubes can be used if mixed with additives that impose a convection-minimal environment. The tubes can be sealed with any material that is inert and does not allow evaporation of the solvent. Most commonly used materials include soft wax, vacuum grease, nail polish.

5.3.1. Single Capillary

1. Place 0.39–11 μL of protein solution on a clean hydrophobic or siliconized surface (clean parafilm sheets work well). Carefully take a quartz or glass capillary tube with both ends opened and touch the protein droplet to allow its uptake by capillary action. We recommend filling the capillaries up to a length of 40 or 60 mm in order to have a wide screening of the phase diagram. The volumes are listed in **Table 5.1**.

Table 5.1. Capillary dimensions and volumes of proteins required.

Diameter	Length		
	40 mm	**50 mm**	**60 mm**
0.1 mm	0.31 μl	0.39 μl	0.47 μl
0.2 mm	1.25 μl	1.57 μl	1.88 μl
0.3 mm	2.82 μl	3.53 μl	4.24 μl
0.4 mm	5.03 μl	6.28 μl	7.53 μl
0.5 mm	7.85 μl	9.81 μl	11.78 μl

One can also load the protein premixed with low-melting agarose to a final concentration of 0.05%. The agarose can be incorporated at low concentration into the crystallization mixture to i) increase the diffusivity of the environment,[4] ii) avoid crystal slipping during the initial quality-screening,[2,7,8] and iii) increase the nucleation rate.[9] The influence of agarose results in a larger number of crystal production providing a wider selection of an optimal quality crystal for X-ray data analysis.

2. Prepare an equal volume of the precipitating salt with the option of including anomalous scattering atoms (such as iodine or bromide) and cryopreservation solution. Carefully layer the precipitant solution against the surface of the protein solution forming a liquid–liquid interface. In our laboratories, iodine or bromide salts are commonly mixed with the precipitating solution including glycerol as the cryoprotectant. We recommend using halide and glycerol concentrations of about 100 mM and 25% (v/v) respectively. The advantages of the simultaneous process of protein crystallization, high X-ray scattering atom incorporation, and

cryoprotection are three fold. First, crystals remain in a stable environment at all times while never being exposed to drastic chemical changes or subjected to physical manipulation. Second, protein crystals can be quickly evaluated by *in-situ* X-ray diffraction from their original growth environment. Finally, since the entire crystallization and incorporation process is done in a closed chamber, the investigator is never exposed to any harmful chemicals after the equilibration process has initiated. Therefore, the entire setup is safe and easy to transport.

3. Seal the ends of the capillary tube with vacuum grease, soft wax, or nail varnish. The final setup is illustrated in **Plate 5.1** and should be rested horizontally in an incubator oven at a selected temperature. During the equilibration process, a supersaturation wave is activated along the protein chamber where the salts initially diffuse into the protein volume allowing nucleation, crystallization, and glycerol treatment to occur in sequential events.

Protein crystals can often be observed within a couple days along the length of the capillary. Even though crystals suitable for X-ray diffraction can be obtained in a few days, at least one week of equilibration time is required for sufficient cryoprotection. The crystals grown in the capillary can range from 0.05 to 4 mm in the longest dimensions where the largest crystals fill the diameter of the capillary.

Single capillaries with diameters of 50 and 100 μm have recently become available and used successfully for crystallization screening with nanoliter volumes of protein.[10] Computer simulation suggests 40-mm-long capillaries with these diameters are sufficient to screen a large window of the solubility diagram (see **Table 5.1**), while reducing the required protein volume to one hundred nanoliters. Viscous forces dominate within these ultrathin capillaries giving rise to an almost complete convective free environment. Triana S&T (www.trianatech.com, last accessed 5/19/2007) offers kits containing PEG and salt precipitants combined with various buffered solutions for rapid screening against proteins filled in these nanoliter capillaries. Once the capillary containing the protein solution is sunk into the precipitating solution made of different PEG and salts, successive wave of supersaturation explores very efficiently the crystallization space.

As the result, a wide crystallization space can be efficiently explored for finding initial or optimized crystallization conditions.

5.3.2. Multiple Array Configurations

Three systems for multiple preparations have been designed and implemented for counter-diffusion crystallization. The first is the Granada Crystallization Box (GCB) commercialized by Hampton Research (www.hamptonresearch.com, last accessed 5/20/2006) and Triana Science & Technology (www.trianatech.com). Six capillary tubes containing proteins are able to simultaneous equilibrate against a common precipitating solution partitioned by a gel (**Plate 5.2**). The technique is referred to as the gel-acupuncture method[11] described below:

1. In a GCG box, a gel containing the precipitating agent is casted. Capillaries are filled with protein solutions and sealed at one end. The capillary is then pierced into the gel layer about 2–3 mm using a guide with holes that is part of the tray.
2. A solution of a buffered precipitating agent (equal volume to the gel volume) is poured onto the gel layer and the GCB tray is covered. Six capillaries are configured for counter-diffusion equilibration.

The second is a crystallization cassette containing 12 multiple capillaries that can be used to uptake many samples of proteins at a time and mixed with several crystallization reagents simultaneously.[3,12] The cassette has a cylindrical design and is made up of two sections that are joined together. The upper section contains multiple capillary tubes that extend downward. The lower section is a disc having multiple slots, with each slot corresponding to a capillary (**Plate 5.3**).

The slots contain the precipitant solution cushioned at the bottom with a wax and sealed with a latex membrane. The protein solution is carefully layered on the membrane. The upper section can be pushed down to contact the protein droplet and consequently by capillary action is pulled into the capillary. Finally the capillary tubes are simultaneously penetrated through the membrane to equilibrate with the precipitating solution.

The third system is a crystallization device that we have designed in collaboration with the laboratory of Dr. L. DeLucas (University of Alabama at Birmingham) to quickly screen or optimize protein crystallization conditions in a 96-array configuration (commercialized by Genomic Solutions). Purified proteins are able to be loaded in capillary trays for finding optimal crystal growth conditions by counter diffusion in a restricted geometry (**Plate 5.4**). The loading of the protein, equilibration process, addition of strong anomalous scattering atoms and, cryogenic soaking can be simultaneously executed.

Ninety-six counter-diffusion processes can be performed at one time in an automated manner. The glass capillaries are immobilized in a rack. The preparation of protein crystallization is shown for one capillary in three phases (**Plate 5.5**). (1) Tray A contains the protein solution prepared on a membrane over a well of precipitant solution. The entire tray is moved to an optimal distance where the protein solution can be drawn into the length of the crystallization chamber by capillary action. (2) Tray B contains soft wax. The wax tray moves towards the other end of the capillary and seals the tube as the capillary penetrates the wax. (3) Tray A continues to move towards the capillary in such way that the tube pierces through the membrane and contacts the precipitating solution. The solution is then allowed to equilibrate by counter diffusion. The protein droplet and precipitating solution are 8 μL and 25 μL respectively. Now the system executes the following steps:

1. The precipitant chamber is filled with the precipitating salts, buffers, or ligands. The protein is loaded into the capillary with the device. The precipitant and protein volumes are allowed to contact each other, resulting in a counter-diffusion equilibration. Protein crystals obtained can be co-crystallized with the studied ligand in the capillary tube for X-ray diffraction.
2. The protein is loaded into the capillary with the ligand and allowed to equilibrate with only the precipitating solution. After equilibration processes by counter diffusion, co-crystallization of the ligand–protein complex is immediately ready for X-ray analysis.
3. The protein solution alone is loaded into the capillary and equilibrated by counter diffusion with the precipitating reagent. After crystals are obtained, a ligand can be sequentially loaded into the reservoir of the precipitating reagents. The ligand is now allowed to slowly diffuse into the protein solution and soak into the already crystallized protein.

5.4. Getting the Protein Crystal to the X-Ray Beam as Soon as Possible

The excitement and joy of discovering a crystal after laborious screening and optimization can be extinguished by many dealings. To determine the usefulness of the crystal obtained in a single droplet from vapor diffusion or batch equilibration, one has to employ invasive actions to transfer the crystalline object into the X-ray beam for diffraction analysis. In most cases, this is no trivial matter since protein crystals (especially those of nucleic acids) are very sensitive to physical manipulation. Whether you are looping your crystal to be cryogenically cooled or transferring them into a capillary tube, the most well-trained steady hands can inadvertently smash, crush, crack, or dissolve the crystal to be analyzed. In many cases, a crystal can be successfully mounted on a loop or in a capillary but the crystal quality is compromised as a consequence of extensive handling. In the case where a crystal is irreproducible or few to obtain for one reason or another, one would be extremely disappointed if the structure determination process stopped here after extensive crystallization efforts.

Obtaining proteins by crystallization in capillaries offer a significant advantage in that there is absolutely no physical handling of the protein crystal necessary for X-ray diffraction analysis. The crystals grown in the capillary can be directly placed in front of an X-ray source while still immobilized in its crystallization volume. This offers a tremendous advantage of quickly diffracting crystals with X-ray *in situ* along the length of the capillary to determine which would be optimal for extensive X-ray data collection for structure determination.

5.4.1. Taking a Walk Down Crystal Lane

When crystals are grown under a supersaturation gradient created by counter diffusion in a capillary, the most notable observation is high nucleation and small crystalline material near the precipitant–protein solution interface. As one move across the length of the capillary fewer but larger crystals should be evident. Under a dissecting microscope, one should visually walk across the length of the capillary and carefully inspect the potentially best diffracting crystals. The regions where there are separate and distinguishable crystal habits should be highlighted with a fine tip permanent marker identifying the region to be aligned and centered to the X-ray beam for preliminary analysis.

5.4.2. Mounting and Aligning Crystals in Capillary

Once the region or regions of interest are identified along the capillary-containing crystals for initial X-ray diffraction, the capillary must be mounted on a goniometer head (**Plate 5.6**). The entire capillary can be simply attached at room temperature on the goniometer multistage platform with model clay or silicone grease such that the capillary is directly perpendicular. The mounting and aligning procedures would be similar to classic techniques where the target crystal is rotated perfectly about a central axis by adjusting the translations. The areas previously identified by the user with markings on the capillary facilitate the general location of the crystals to be analyzed. Usually the X-ray apparatus with a horizontal phi spindle is ideal for mounting capillaries because it minimizes the chances of crystal slippage. However, small diameter capillaries (smaller than 0.3 mm) or large diameter capillaries with gelled proteins can be mounted straight either to the vertical or horizontal phi spindles.

5.4.3. Preliminary X-Ray Analysis

Still or oscillation photographs can be taken for targeted crystals. Usually to determine which crystal would be the best for complete data collection, the targeted crystal should be diffracted at 0 and 90 degrees for 3–5 minutes at a home X-ray source and 5–10 seconds at a synchrotron source. The distance of the detector can be adjusted according to the observed diffraction limit. Usually a distance that can measure diffraction spots to 3.0 Å is recommended as a good starting point. The first diffraction images should reveal at least three important parameters to consider; the resolution limit, the mosaicity of the reflection spots, and how well the reflections can be indexed.

The estimated resolution limit can be determined by examining how far the reflection spots are observed in the most outer edge with intensity over noise ratio of at least 3. Most data-collection software such as HKL 2000[13] or MOSFLM[14] can reveal exactly what resolution is each reflection spot. Generally, if one can measure the furthest spot diffracting away from the detector center the resolution limit can be calculated by the following relation $\lambda / (2\sin((\tan^{-1}(y/x))/2))$ where λ is the wavelength of the X-ray used, y is the distance of the diffraction spot from the center, and x is the distance between the crystal and the detector. It is best to draw a general circle in the diffraction image on the outermost circumference where diffraction spots are observed. The resolution limit should reveal many reflection points, not just one isolated diffraction spot. The detector can be moved in towards the crystal to obtain higher resolution diffraction and moved away to obtain lower resolution spots at the outer edge.

The tightness of spot corresponds to the crystal's mosaicity or internal lattice order. The tighter and more round the spots are the less mosaic it is. Highly mosaic crystals will have diffraction spots that may appear elliptical or have a cigar shape. Usually, crystals with low mosaicity will have circular and well-separated reflections.

The identification of each reflection spots to crystallographic indices and accurate distance measurements between the reflections are required for the proper assignment of the crystal's unit cell dimensions and space group. The details and principles of these practices will not be discussed here but is refereed to Stout and Jensen[15] for a more comprehensive description of these crystallographic techniques. The important point here is that the crystal to be used for structure determination should display diffraction spots that are easily separated such that their location and intensity can be easily measured for subsequent calculations for structure determination.

In the practical sense discussed here, crystal quality should be determined by its resolution limit, mosaicity, and its ability to have its reflection position and intensity identified and measured respectfully. The crystals with the best quality should be chosen for complete data collection for subsequent structure determination. Each crystal along the capillary within the selected area should undergo preliminary X-ray analysis and the one evaluated to be the best crystal in terms of what has been discussed so far should proceed to complete data collection. We have observed and demonstrated that crystals along the span of the capillary do not have equivalent quality. Crystals grown along the capillary length will experience diverse supersaturation environments and thus will be of different worth.

5.4.4. Data Collection and Electron Density Calculation

When the crystal (or crystals) of choice has been chosen, the best data collection strategy must be considered. The factors that should be considered include the temperature of the data collection, amount of the data to collect, and method of crystallographic phasing to obtain an initial electron density map.

5.4.4.1. To Freeze or not to Freeze; That Is the Question

Let us reflect on whether the data collection should be performed under room temperature or cryogenic conditions. There are advantages and disadvantages for both conditions. Data collection at room temperature would not require the crystal to be grown or treated with cryogenic preservatives. Furthermore, the capillary with its crystal do not need to be quickly flash cooled and as the result decreases the chances of rendering the crystal to be more mosaic. The disadvantage, of course, is that crystals exposed to X-ray at room temperature decay much faster and often does not last through a complete data set. In this respect, synchrotron data collection would be almost impossible to perform at room temperature due to the immense X-ray beam intensity. Our personal preference is to undergo data collection under non-cryo conditions at a home source when possible because it is less invasive and eliminates a step where a crystal can be destroyed. If a protein crystal is too sensitive for X-ray analysis, several crystals will be required for a complete data set. However, crystals destined for synchrotron X-ray analysis must be prepared for cryogenic data collection.

As mentioned above, the preparation of protein crystals for cryogenic data collection can be prepared by allowing the crystals to grow against a precipitating agent containing a cryogenic preservative already or by the addition of a cryogenic solution in the precipitating solution after crystals are formed. In either case, the capillary can be directly flash frozen without significant damage to the crystal or capillary without ice that build up on the targeted area. If the capillary tube is dirty, be sure to clean gently the surface with an ethanol or equivalent solution to avoid any nucleation sources for ice buildup. It is often the case, where the region of the capillary containing the targeted crystals are excised with a razor blade and remounted on the goniometer with model clay. No resealing is really necessary since the capillary will be frozen. The primary reason for truncating the capillary tube is to minimize the vibration of the crystal caused by the cryostream. This is especially relevant for capillary tubes with diameters of 0.1 mm or less.

5.4.4.2. Processing Data to Phasing for *de Novo* Structure Determination

Data collection and subsequent data reduction can be performed on crystals in capillary by conventional methods. While there are many methods to process data and calculate phases for structure determination, we shall outline a strategy for the most rapid method for *de novo* structure determination by Single-wavelength Anomalous Scattering (SAS). Crystallographic explanations and details of these procedures have been reported by Ng et al.[3] The details of the crystallographic methods involved in determining a protein structure will not be explained or discussed here. We only draw attention to some important steps that you or with the help of your favorite crystallographer can conduct. **Plate 5.7** highlights three possible schemes using different publicly available software packages to perform the following tasks:

1. Record data set over a single crystal with 9–12 fold redundancy. Select a rotation range to yield high-redundant anomalous data without employing any inverse beam technique. Determine space group, unit cell parameter, and number of mol per asymmetric unit.
2. Examine data collection statistics and make sure R_{merge} or R_{symm} values, completeness, and Wilson plots are satisfactory.
3. Calculate Bijvoet difference Patterson maps to identify any significant anomalous signals. Identify and refine strong scattering atom positions.
4. Generate initial phases and their respective anomalous difference Fourier (ADF) maps to find other strong electron density peaks corresponding to additional heavy atom sites.
5. Calculate electron density maps with known calculated phases and search for clear definition of solvent region and degree of connectivity for unambiguous model tracing.

5.5. Important Points

1. Initial screening and optimizing solubility conditions can be performed using the counter-diffusion technique for protein (or any other

macromolecules such as nucleic acids, carbohydrates, viruses, fatty acids, and small molecules of biological significance) crystallization.

2. In comparison to vapor-diffusion or batch methods, the region of the crystallization space screened by counter diffusion is much larger for a single experiment.
3. No image storage is required because the results of the different crystals obtained in each experiment are "fossilized" along the capillary.
4. Because crystallization inside the capillary occurs at conditions progressively approaching equilibrium, selecting crystals suitable for structure determination is not a matter of luck but can be systematically evaluated by X-ray analysis *in situ*: the last grown crystals in the upper part of the capillary are of higher quality.
5. Crystals grown in the capillary by counter diffusion can be used for direct X-ray data collection, phasing by various crystallographic methods and calculation of an initial electron density map without ever touching the crystal.
6. Ligand or drug complexes can be screened for crystallization and used for subsequent structure determination.
7. Crystallization conditions for large sample number of proteins can be screened or optimized quickly and efficiently making the counter-diffusion method very useful for high-throughput applications.

References

1. Garcia-Ruiz JM. **Counterdiffusion methods for macromolecular crystallization**. *Meth Enzymol* 2003, **368**:130–154.

2. Gavira JA, Lopéz-Jaramillo J, Toh D, Garcia-Ruiz JM, and Ng JD. ***Ab initio*** **crystallographic structure determination of insulin form protein to electron density without crystal handling**. *Acta Cryst* 2002, **D58**:1085–1254.

3. Ng JD, Gavira JA, and Garcia-Ruiz MA. **Protein crystallization by capillary counter-diffusion for applied crystallographic structure determination**. *J Struct Biol* 2003, **142**:218–231.

4. Garcia-Ruiz JM, Otalora F, Novella ML, Gavira JA, Sauter C, and Vidal O. **A supersaturation wave of protein crystallization**. *J Cryst Growth* 2001, **232**:149–155.

5. Salemme FR. **Protein crystallization by free interface diffusion**. *Meth Enzymol* 1985, **114**:140–141.

6. Kuta Smatanova I, Vacha F, Gavira JA, and García-Ruiz JM. **Photosystem II core complex of *Pisum sativum* crystallization trials.** *Material Structure* 2004, **11**:30–31.

7. López-Jaramillo FJ, García-Ruiz JM, Gavira JA, and Otálora F. **Crystallization and cryocrystallography inside X-ray capillaries**. *J Appl Cryst* 2001, **34**:365–370.

8. Biertümpfel C, Basquin J, Suck D, and Sauter C. **Crystallization of biological macromolecules using agarose gel.** *Acta Cryst* 2002, **D58**:1657–1659.

9. Vidal O, Robert MC, and Boué F. **Gel growth of lysozyme crystals studied by small angle neutron scattering: Case of agarose gel, nucleation promotor.** *J Cryst Growth* 1998, **190**:257–270.

10. Garcia-Ruiz JM and Gonzalez-Ramirez LA. **Procedure for screening crystallization conditions of biological macromolecules. [Procedimiento para el chequeo de condiciones de cristalización de macromoléculas biológicas.]** Spanish Patent 2006; PCT/ES/2006/070004.

11. Garcia-Ruiz JM. **Uses of crystal growth in gels and other diffusing-reacting systems.** *Key Engineering Materials* 1991, **88**:87–106.

12. Pusey ML, Liu Z-J, Tempel W, Praissman J, Lin D, Wang B-C, Gavira JA, and Ng JD. **Life in the fast lane for protein crystallization and X-ray crystallography**. *Progress in Biophysics and Molecular Biology* 2005, **88**:359–386.

13. Otwinowski Z and Minor W. **Processing of X-ray diffraction data collected in oscillation mode.** *Meth Enzymol* 1997, **276**:307–326.

14. Leslie AGW. **Automated data collection and processing for macromolecular crystallography.** *In:* ***Joint*** **CCP4 + ESF-EAMCB** *Newsletter on* ***Protein Crystallography*** 1992, No. 26.

15. Stout GH and Jensen LH. ***X-Ray Structure Determination: A Practical Guide,*** 2nd ed. New York: John Wiley and Sons Publication, 1989.

16. Collaborative Computational Project 4. **The CCP4 Suite: Programs for protein crystallography**. *Acta Cryst* 1994, **D50**:760–763.

17. Jones TA, Zou JY, Cowan SW, and Kjeldgaard M. **Improved methods for building protein models in electron density maps and the location of errors in these models**. *Acta Cryst* 1991, **A47**:110–119.

18. McRee DE. ***Practical Protein Crystallography***, 2nd ed. San Diego, California: Academic Press, 1999.

19. Wang BC. **Resolution of phase ambiguity in macromolecular crystallography.** *Meth Enzymol 1*985, **115**:90–112.

6

Scale-Down Approaches for Measuring Protein-Protein Interactions

Joseph C. Fanguy, Charles S. Henry, Steven C. Holman, Joseph J. Valente, and W. William Wilson

Abstract

The osmotic second virial coefficient, *B*, is a thermodynamic parameter that has been shown to have predictive properties relevant to protein crystallization, protein solubility behavior and protein colloidal stability. Solution conditions favorable for crystallization have *B* values that are slightly to moderately negative, indicating a mild net attraction between protein molecules. The dependence of *B* on various solution parameters such as pH, temperature, and ionic strength has been shown to mimic the solubility dependence on the same parameters so that solubility behavior can be estimated quickly from dilute solution measurements for *B*. Predicting the propensity of a protein to aggregate in a given solvent is important in protein formulation, and *B* values that are positive, indicating net repulsion between protein molecules, can be used as a predictor for colloidal stability. Although *B* is an attractive predictive parameter for these applications, its utilization as a screening parameter has not yet been fully realized because of protein sample and experiment time requirements. We report here two methods that may make it more feasible to incorporate *B* measurements into a screening strategy. One is a modified static light-scattering approach utilizing simultaneous detection of scattered intensity and protein concentration while the other is a microscale version of self-interaction chromatography. Both methods significantly reduce sample and time resources per *B* determination.

6

Scale-Down Approaches for Measuring Protein-Protein Interactions

Joseph C. Fanguy,[*] Charles S. Henry,[†] Steven C. Holman,[*] Joseph J. Valente,[†] and W. William Wilson[*]

[*]*Mississippi State University, Mississippi, USA*
[†]*Colorado State University, Fort Collins, Colorado, USA*

6.1. Introduction

The human proteome dictated by about 30,000 genes will yield an avalanche of 100,000 to 300,000 proteins,[1] most with unknown structure and function. X-ray crystallography is currently the workhorse for determining the three-dimensional structure of proteins; however, many of these "new" proteins will defy crystallization using traditional trial-and-error screening methods. An international summary of results from 17 structural genomics centers shows that, to date, only about 35% of the purified water-soluble proteins have produced crystals, and of those, only about half have led to diffracting crystals. About a third of the genome codes for membrane proteins and nearly 70% of all drugs target them, but crystallization of membrane proteins presents unique challenges. Selection of detergent type and concentration along with limited protein quantity are just two of the typical constraints when conducting crystallization trials. Additional variables, which may influence crystallization of both soluble and membrane proteins, include buffer type and pH, temperature, crystallizing agent and concentration, protein concentration and protein stability. Current technology has resulted in high-throughput fluidics (liquid handling) and crystallization systems such that it is possible to prepare thousands or even tens of thousands of crystallization trials. However, the sheer number of trials is itself not a recipe for success, as evidenced by a relatively low success rate. It will be incumbent upon researchers to develop a different, more fundamental strategy that will improve the probability for successful crystallization for "difficult" proteins.

Formulation development of protein pharmaceuticals has led to promising treatments of human diseases and conditions (diabetes, cancer, and hemophilia), but stability of the protein formulations is a critical issue for long-term efficacy. For economic viability, the shelf life required for a typical protein pharmaceutical is about 24 months.[2] Proteins undergo both chemical (oxidation and deamination) and physical (denaturation, partial unfolding, and aggregation) degradation. Protein formulations are often administered at doses with concentrations exceeding 100 mg/mL, and the approach to achieve these high-concentration formulations has, so far, been quite empirical. For liquid formulations, thermodynamic stabilization is the most important consideration for preventing aggregation of the protein during storage.[3] Stabilization is often attempted by the addition of a solute, which will impose a large excluded volume.[4] These solutes, sometimes referred to as excipients in pharmaceutical terms, include most amino acids, sugars, polyols, and polymers. While these compounds generally increase the free energy required to unfold the protein, they may also lower the solubility appreciably, requiring a careful balancing act.[4]

The crystallization of proteins and the stabilization of protein pharmaceutical formulations have a commonality in that both processes are inextricably linked to the thermodynamics of protein aggregation. In the case of crystallization, aggregation is prerequisite for the formation of a nucleus; a critical sized ordered aggregate capable of sustaining ordered growth. Conversely, colloidal stability of pharmaceutical formulations generally depends on the suppression of aggregation allowing efficacy to be maintained. So, in either case, it is ultimately the protein-protein interactions in the specified solvent condition that dictates the process outcome. For example, recent work by Chi et al.[5] demonstrated that for granulocyte colony-stimulating factor (G-CSF), the magnitude of the protein-protein interactions as measured by a thermodynamic property called the second virial coefficient, B, for various sucrose (excipient) concentrations roughly correlated with aggregation rate, as well as with the denaturation temperature. In general, more positive B values, indicating more repulsive interactions, produced slower aggregation rates.

The role of protein-protein interactions regarding crystallization as estimated by the second virial coefficient is well established both experimentally and theoretically.[6–13] The finding is that a range of values of the second virial coefficient, B, is correlated to solution conditions that are favorable to crystallization. A "crystallization slot" corresponds to B values in the range of about -1×10^{-4} to -8×10^{-4} ($mol \cdot mL \cdot g^{-2}$) and indicates protein-protein interactions that are slightly or moderately attractive. Protein crystal growth (PCG) experiments conducted in solution conditions at more negative B values have a greater risk of forming amorphous solid phase because of corresponding stronger protein-protein attractions. On the other hand, experiments at more positive values, where the net protein-protein interactions are repulsive, typically require protein concentrations that are impractically high to cause phase separation of any kind. The crystallization slot can be used as an effective guide by PCG'ers to direct changes in a particular solution parameter (pH, temperature, crystallization agent concentration, etc.) that will increase the probability of a successful crystallization trial. It is important to point out that conducting PCG experiments under conditions that correspond to the crystallization slot does not guarantee a successful crystallization trial. However, working at conditions outside the slot most assuredly reduces the probability of a desirable outcome.

Another aspect of protein-protein interactions that directly impacts both crystallization and formulation stability is the influence on protein solubility. Protein solutions used to measure B values are typically well undersaturated, usually having concentrations in the 1–5 mg/mL range. However, results presented by Muschol and Rosenberger[14] for lysozyme with NaCl as the crystallizing agent showed that the protein-protein pair potentials manifest in undersaturated (even dilute) protein solutions extend into supersaturation regions. Experimental data correlating B with about 25 solubility (s) values for lysozyme in various crystallizing agents along with additional data for ovalbumin, ESA, and for lysozyme in H_2O and D_2O have been reported.[10,15–17] The empirical correlation between B (a dilute solution parameter) and s (a phase transition parameter) was intuitive at best. After all, solubility depends on the binding energy between closely spaced protein molecules having very specific mutual orientations in the crystal lattice. By contrast, the second virial coefficient is a statistical average of interactions between two protein molecules in the liquid phase accounting for all relative separation distances and orientations, with each configuration weighted by a Boltzmann factor. Nevertheless, the experimental data demanded that a simple theoretical relationship between B and s must exist, and subsequent publications[18,19] provided the theoretical basis for the direct link between B and s. Major findings of these studies were that protein interactions were strongly anisotropic and that crystallization conditions had little effect on the interaction distance or the anisotropy between the protein molecules.

The correlation between B and s offers the practitioners of protein crystal growth and formulation development a distinct advantage for determining the solubility behavior of proteins. From the practical standpoint of setting up PCG or formulation trials, it is desirable to know how a protein's solubility depends on a particular solution parameter. For example, for any rational design of trials, it is important to know if and to what extent the protein's solubility has normal, retrograde, or no temperature dependence in a given solvent. In other cases, knowledge of the dependence of s on solution pH, ionic strength, type, and concentration of crystallization agent or excipient, etc. will allow systematic trials to be conducted. Since it is impractical to determine absolute values for s by equilibrium studies as a function of so many solution variables, the measurement of B provides a realistic alternative for obtaining solubility behavior.

Historically, measurement of the second virial coefficient for macromolecular solutions has been performed primarily by osmotic pressure,[20] sedimentation equilibrium,[21] or static light-scattering (SLS)[22] experiments. For purposes of a screening technology, osmotic pressure and sedimentation equilibrium are not practical, both from an experiment time and sample quantity point of view. Traditional SLS has been the most widely reported method for B determinations during the past decade in spite of its labor and sample intensive requirements. Although the framework-relating B to the protein-protein interactions so significant in the processes of crystallization and formulation stability has been well established, its utilization as a primary screening parameter for these processes has not been realized. To achieve this goal, a way must be found to measure B that is relatively fast (on the order of minutes per B determination), requires small amounts of protein (microgram or submicrogram quantities per B determination) and lends itself to a robotic platform that is consistent with a high-throughput system. In the following pages, we describe two scale-down approaches for achieving this goal that are significant departures from the conventional methodology for measuring protein-protein interactions as reflected by the second virial coefficient.

6.2. Scale Down of Static Light Scattering

6.2.1. Configuration Description

The traditional SLS experiment requires measurement of the scattered light intensity from a protein solution in excess of background as a function of protein concentration. The usual procedure is (1) to prepare four to eight dilutions of a protein stock solution spanning a concentration range of approximately 1–5 mg/mL (depending on the molecular weight and solubility of the protein), (2) to filter the solution into the scattering cell, and (3) to measure the scattered intensity. The concentrations of the protein solutions are usually determined by absorption spectrophotometry as separate experiments. Filtration of the protein solutions is necessary to remove or at least minimize the ubiquitous dust particles in aqueous solutions. These preliminary procedures require time and consume precious protein. We have modified the traditional SLS experiment in two important ways in order to minimize both time and protein required for a single *B* measurement. The basic features of the scale-down SLS (SD/SLS) experiment are shown in **Fig. 6.1**. The first modification is the incorporation of a low-volume (approximately 1 μL) scattering cell described later. The second modification is a configuration allowing the simultaneous measurement of scattering and protein concentration. In short, this is accomplished by using a bifurcated fiber to deliver both the incident laser beam for scattering and the incident UV beam for absorption (protein concentration) measurements. The utilization of this dual detection configuration was presented by Wilson,[23] and a recent report[24] described a modification of a commercially available light-scattering instrument to perform a similar function but using a significantly different optical configuration.

The intent of the configuration shown in **Fig. 6.1** is to allow the simultaneous measurement of light-scattering intensity and protein concentration

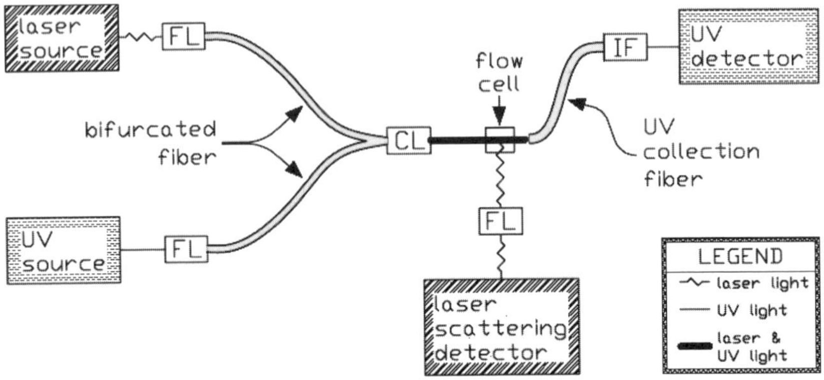

Figure 6.1. Schematic of the scale-down SLS experiment for simultaneous detection of light-scattering intensity and protein concentration.

that will allow the determination of the second virial coefficient from a single injection of protein sample into the flow system. One crucial aspect of this approach is the delivery of both UV and laser light into the sample cell channel. This was accomplished using a bifurcated fiber assembly, which contains two fibers that are packaged as separate arms on one end and side-by-side arms on the other (common) end. This design permits the coupling of two different light sources to each of the separate fiber legs with the common end of the fiber transmitting both light beams into the sample cell. The assembly used for this experiment was an Ocean Optics (Dunedin, Florida) QBIF200-UV-VIS, which was comprised of 200 μm quartz (UV) and 200 μm silica (visible) fiber. A 50-mW laser diode (Uniphase) provided incident light at 532 nm, which was coupled into the silica optical fiber via an OZ (Ottawa, Canada) fiber optic source coupler. The OZ coupler uses a focusing lens (FL) to focus a collimated light beam to a spot size matching the core of an optical fiber. The other (quartz) leg of the bifurcated fiber was coupled to a 69050 Spectral Luminator (Oriel Instruments, Stratford, Connecticut), which served as the UV light source. This illumination unit contained an arc lamp and grating monochromator that allowed the output wavelength to be tuned from the UV to the visible region of the spectrum. To optimize coupling efficiency of UV light into the fiber, another focusing lens (Ocean Optics, Dunedin, Florida) was connected to the end of the SMA-terminated quartz fiber. This optical component collected and focused the beam of light from the UV illumination source onto the tip of the fiber. The two output beams (532 nm VIS and 280 nm UV) passed through a collimating lens (CL) into the sample flow cell, which is described in **Fig. 6.2**.

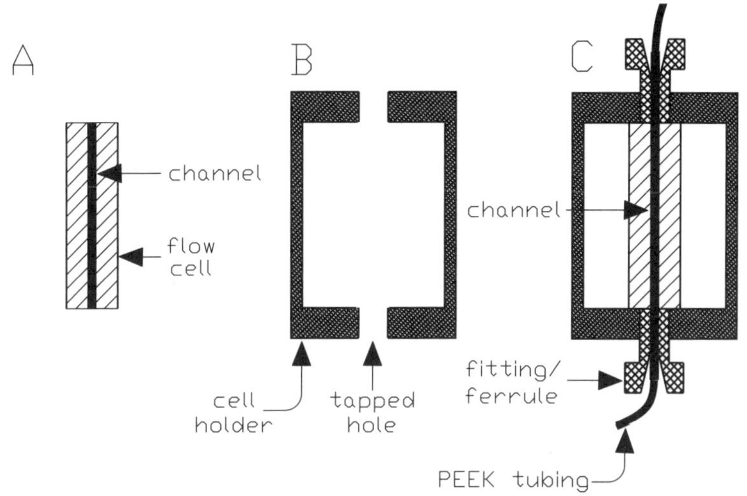

Figure 6.2. Sketches of (**a**) low-volume cell; (**b**) cell holder; (**c**) cell holder, cell, and PEEK fittings.

The channel of a quartz cytometery cell (Hellma, Plainview, New York) served as the flow path of buffer/sample solutions for the scattering/absorption experiments. The rectangular cell was 4.2 × 4.2 × 20.3 mm with a 0.25 × 0.25 mm channel that extended through the middle of the entire length (20.3 mm) of the cell (**Fig. 6.2a**). Although this small inner diameter channel design (total volume of 1.3 μL) was ideal for low-volume injections, the configuration provided no means for directly connecting inlet/outlet tubing to the channel. Therefore a cell holder (**Fig. 6.2b**) was designed and constructed of aluminum to mechanically stabilize the cell for mounting and to permit a sealed connection between the inlet/outlet tubing and the channel. The cell holder contained an access window (for entrance/exit of light beams) on each side. In addition, a 1/4-28 hole was tapped in both the bottom and top of the holder for connecting the appropriate fitting/tubing to the cell (**Fig. 6.2c**).

The transmitted UV signal was collected using a 1-mm core quartz optical fiber that was mounted approximately 1 cm away from the face of the cell. The transmitted UV light was delivered to a solid-state detection source, a HC220-01 silicon photodiode/amplifier (Hamamatsu Corporation). A 280-nm interference filter (IF) from Melles Griot was mounted directly before the UV detector (**Fig. 6.1**) to block extraneous light from reaching the UV detector. A second solid state detector (Viscotek) was positioned to collect the light scattered at 90° to the incident beam. A collection/focusing lens (**Fig. 6.1**) was used to collect the scattered light and focus the magnified (4×) image onto a 0.3-mm-diameter pinhole mounted just before the detector surface.

6.2.2. Flow Injection Components

The major system requirement for performing any type of online analysis is a constant flowing solvent stream into which plugs of sample can be introduced reproducibly. In addition, the cleanliness of both buffer and sample solutions is especially important when laser-light scattering is utilized as a method of detection for an online method. The organization of flow-injection components and filters used as part of the online methodology for determining the second virial coefficient are shown schematically in **Fig. 6.3**.

The 8-mL stainless-steel syringe (Harvard Apparatus, Holliston, Massachusetts) contained the buffer in which the light-scattering experiment was performed. In preparation for an online second virial coefficient measurement, the syringe was rinsed with a few milliliters of the appropriate buffer and then loaded with 2.5 mL of fresh solvent. The syringe was then mounted in a Harvard Apparatus PHD 4400 programmable syringe pump, which maintained a constant flow rate of solvent during the experiment. Upon establishing a stable flow of buffer (40 μL/min during experimental setup), the remaining components of the flow injection analysis system were incorporated in the order specified below.

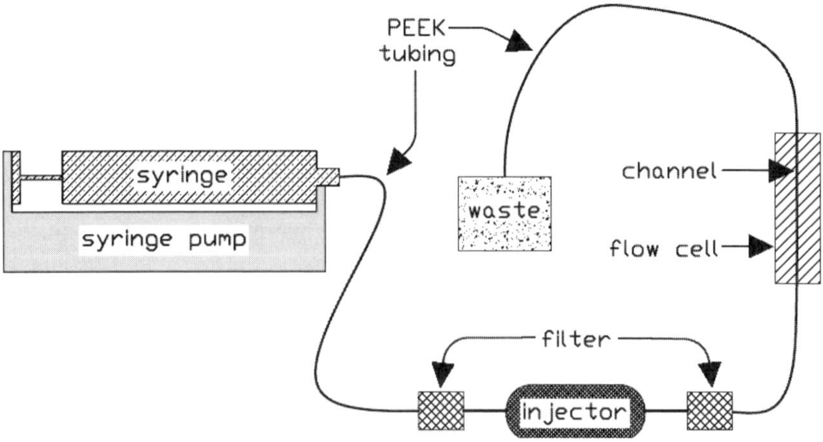

Figure 6.3. Schematic of components for the flow injection system.

The first inline stainless steel filter (Upchurch Scientific, Oak Harbor, Washington) contained a 0.5-μm stainless-steel frit followed by a 0.2-μm Nuclepore Track-Etch Membrane filter (Whatman, Florham Park, New Jersey) and was connected to the syringe via 15 cm of 0.5 mm ID PEEK tubing. This filter ensured that clean buffer solution reached the injector (Model 7413, Rheodyne, Rohnert Park, California) where the sample was introduced into the system. The model 7413 is designed for low-volume injections as it contains an internal loop disk to which three interchangeable sample loops of volumes 0.5, 1, and 5 μL are attached. A length of about 10 cm of 0.25 mm ID PEEK tubing was used between the sample injector and the flow cell in an effort to maintain the integrity of the sample plug. The shorter length of tubing decreased the time required for the injected sample plug to reach the detector, thereby minimizing the amount of diffusion before detection. Secondly, the 0.25-mm inner-diameter tubing matched the channel dimension of the detection cell and as result minimized turbulent mixing that would have occurred with mismatched diameters. In addition to preservation of the plug, sample cleanliness was maintained online by placing another stainless steel filter between the injector and detection cell. This filter was configured in the same manner as the first and removed particulate matter introduced into the system during sample injection.

6.2.3. Data Acquisition System

The outputs of both the UV and laser scattering detectors were in the form of a 0–10-V DC signal. Therefore, simultaneous acquisition of these signals was accomplished using a National Instruments 12-bit PCI-6024E board, which featured 16 channels of analog input. A LabView program was written to read the data acquisition board, process the signal, and then output the data in a spreadsheet format. A data acquisition rate of 1 sample/second was utilized for all experiments, unless stated otherwise.

6.2.4. Samples for Demonstration Purposes

Two proteins and several solvent conditions were used to demonstrate the utility of an online method for determining the second virial coefficient. The procedures used to prepare these sample and buffer solutions are described here.

Calcium chloride (laboratory grade), sodium chloride (certified ACS), sodium hydroxide and 4-(2-Hydroxyethyl)-1-Piperazineethanesulfonic Acid (HEPES) (enzyme grade) were obtained from Fisher (Fair Land, New Jersey). Hen egg white lysozyme (6× crystallized) (Lot # E02301) was purchased from Seikagaku Corporation (Falmouth, Massachusetts), and Concanavalin A (Lot # B49714) was obtained from Calbiochem (La Jolla, California). Glacial acetic acid (17.4 M, reagent ACS) was obtained from Fisher. All chemicals were used as received.

A 0.1 M acetic acid/sodium acetate buffer was prepared by dissolving 6.0 g of glacial acetic acid in approximately 900 mL of distilled/deionized water (Milli-Q Academic, Billerica, Massachusetts). This solution was titrated to a pH of 4.2 using 1 M NaOH and then diluted to 1000 mL with distilled/deionized water. The 0.1 M acetic acid/sodium acetate buffers containing 2% and 5% (w/v) sodium chloride were prepared in the same manner as above, with the exception that 20 g and 50 g, respectively, of sodium chloride were dissolved along with the glacial acetic acid. For lysozyme sample preparation, the protein was slowly dissolved into the appropriate 0.1 M HAc-NaAc buffer. The final concentration of the stock solution was then determined spectrophotometrically (Beckman DU 640 Spectrophotometer) at 280 nm using A (1% (w/v), 1 cm) = 26.3.[25]

A 0.01 M HEPES buffer with 100 μM $CaCl_2$ was prepared by dissolving 0.12 g of HEPES and 0.55 mg of $CaCl_2$ in about 40 mL of distilled/deionized water. The solution pH was adjusted to 7.8 by titrating with 1 M NaOH and then diluted to 50 mL with distilled/deionized water. Concanavalin A was slowly dissolved (without agitation) into the HEPES buffer to prepare a stock solution whose concentration was determined spectrophotometrically at 280 nm using A (1% (w/v), 1 cm) = 13.0.[25]

All buffer solutions were stored at room temperature and used within a two month period of the preparation date. In addition, all buffer and sample solutions were manually filtered through a 0.2-μm Anotop 10 inorganic membrane filter (Whatman, Florham Park, New Jersey) before use. All stock protein solutions were used within 24 h of preparation for both concentration determination and measurement of the second virial coefficient.

6.2.5. Results for Demonstration Proteins

In order to show the utility of simultaneous detection of scattering intensity and protein concentration for *B* measurements by SLS, a weakly scattering protein (lysozyme) and a stronger scattering protein (Concanavalin A) were chosen for study. The intensity and concentration data were treated according to the SLS working equation[22]

$$K_C/R_{90} = 1/M + 2Bc, \tag{6.1}$$

where K is an optical constant (cm^2 mol g^{-2}) given by $K = 4\pi^2(dn/dc)^2 n_0^2 / (N_A \lambda^4)$, c is the protein concentration (g cm^{-3}), R_{90} is the Rayleigh factor (cm^{-1}) at angle 90°, M is the molecular weight of the protein (g mol^{-1}), B is the second virial coefficient (mol mL g^{-2}), dn/dc is the refractive index increment (cm^3 g^{-1}), n_0 is the solvent refractive index, N_A is Avogadro's number (mol^{-1}), and λ is the wavelength (cm) of the incident light in a vacuum. According to **Eq. 6.1**, a plot of Kc/R_{90} vs. c (often called a single-angle Zimm plot) linearizes the SLS data, and B is determined from the limiting slope.

For comparison purposes, one data set is shown for the conventional SLS experiment in which four discrete concentrations of protein were used for the B determination. **Fig. 6.4** displays the typical light-scattering signal as a function of protein concentration obtained from a conventional flow system experiment.

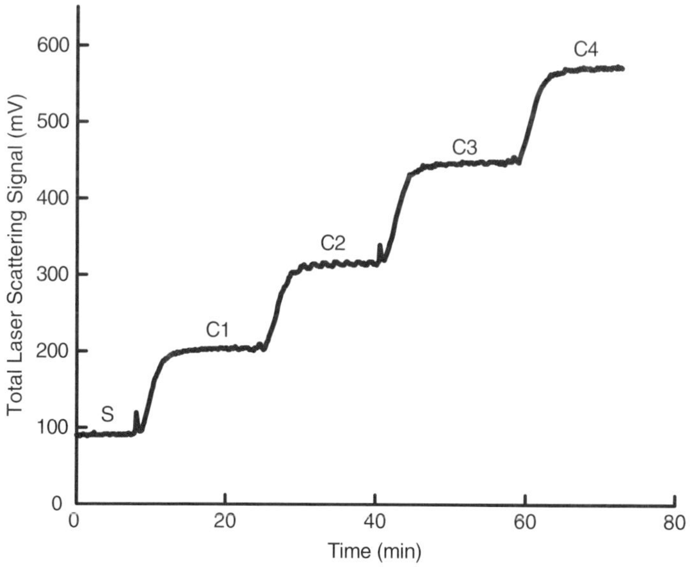

Figure 6.4. Typical trace of a conventional SLS signal in a flow system. See text for details.

Briefly, 200 μL of each of four different concentrations of lysozyme solutions (dissolved in 0.1 M acetic acid/sodium acetate buffer with 5% w/v NaCl) were sequentially introduced into a flowing stream of solvent that passed through the detection cell of a Dawn F unit from Wyatt Technology. The light-scattering signal from the solvent is indicated by S and for the four protein concentrations as C1–C4. The signals were converted to R_{90} values and the data cast according to **Eq. 6.1** to generate the results shown in **Fig. 6.5**.

For the 5% (w/v) NaCl solvent corresponding to the data set shown in **Fig. 6.4**, the *B* value was -5.2×10^{-4} mol mL g^{-2}. Similar experiments were conducted at 0% (w/v) NaCl and 2% (w/v) NaCl and the corresponding values for *B* are indicated in **Fig. 6.5**. Two points of consideration for these conventional experiments are the nearly 80 minutes required for each of the scattering measurements and the approximately 2 mg of protein utilized for each solvent condition.

In a contrasting experiment, the same protein in the same solution conditions was subjected to SD/SLS. **Fig. 6.6** shows the UV and SD/SLS detector responses for a 5 μL injection of 5 mg/mL lysozyme dissolved

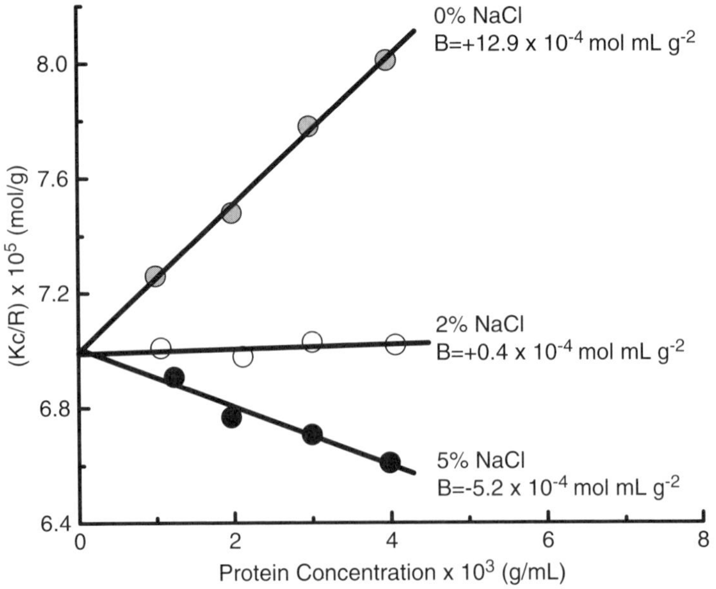

Figure 6.5. Single-angle Zimm plots generated from conventional SLS data as shown in **Fig. 6.4**.

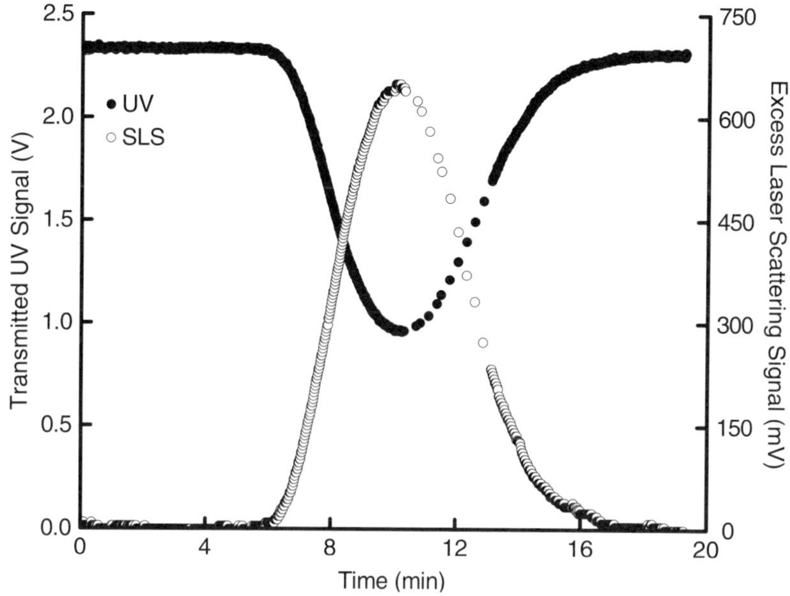

Figure 6.6. Dual traces depicting the simultaneous detection of excess (solvent subtracted) scattering and transmitted UV signals from a single injection of protein solution.

in 0.1 M acetic acid/sodium acetate buffer with 5% w/v NaCl. As the protein solution began to pass through the quartz cell at a flow rate of 1 μL/min, the transmittance of the UV beam began to decrease while the light-scattering intensity simultaneously began to increase. A smooth trace for both signals is evident.

A portion of each trace on the backside of the peak is presented with every fourth data point for visual clarity of point alignment. The individual transmittance values were converted to absorbance and ultimately to protein concentration values. The single-angle Zimm plot from the SD/SLS data was generated and is shown in **Fig. 6.7**. For the 5% (w/v) NaCl solvent condition, a B value of -4.9×10^{-4} mol mL g^{-2} compares favorably with that from the conventional experiment shown in **Fig. 6.5**. The B values for the two other solvent conditions are in quite good agreement with their counterparts shown in **Fig. 6.5**.

To demonstrate the applicability of the SD/SLS system for analyzing higher molecular weight proteins, the well characterized plant sugar-binding protein Concanavalin A (tetramer of molecular weight 104,000 Da) was evaluated in a nontraditional HEPES buffer using both the SD/SLS and

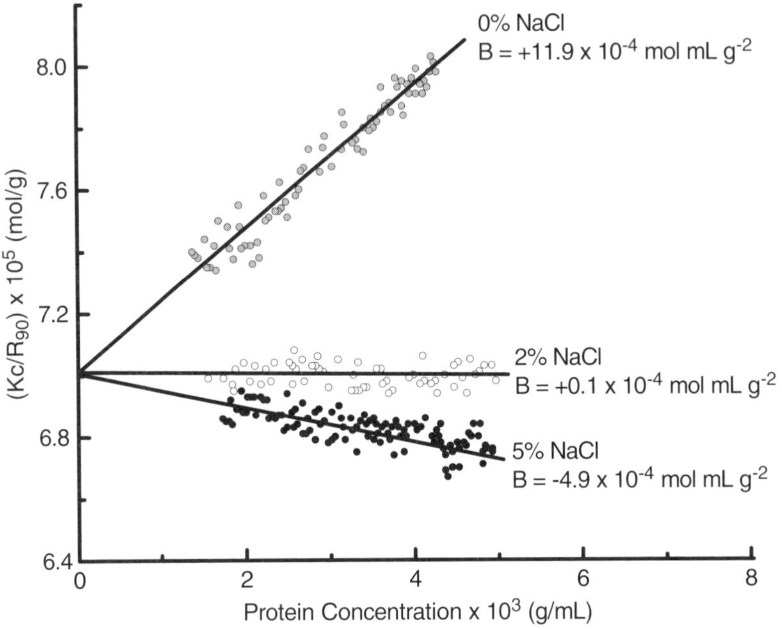

Figure 6.7. Single-angle Zimm plots generated from SD/SLS data similar to that shown in **Fig. 6.6** for lysozyme in three different solution conditions.

the conventional SLS technique. Two separate runs using the conventional SLS approach (maximum protein solution concentration of 2.7 mg/mL) resulted in B values of $+2.4 \times 10^{-4}$ mol mL g^{-2} and $+1.9 \times 10^{-4}$ mol mL g^{-2}. In comparison, the SD/SLS approach (maximum concentration of approximately 2.6 mg/mL detected) resulted in an average B value of $2.2 \pm 0.2 \times 10^{-4}$ mol mL g^{-2} ($n = 2$). An overlay of the single-angle Zimm plots resulting from both techniques is shown in **Fig. 6.8**, illustrating excellent agreement between the data points and corresponding B values.

In terms of operational performance when compared to conventional SLS, the SD/SLS experiment required significantly less time per B determination. The time savings is not only realized during the actual scattering measurements but also for sample preparation since no carefully prepared separate dilutions are required. More importantly, the amount of protein required for a single B measurement has been dramatically reduced to about 25 μg for the SD/SLS compared to about 2 mg for the corresponding conventional SLS. Another advantage of the SD/SLS evident from **Fig. 6.7** is that the data density has been increased on the order of 100 fold, which leads to improved statistical accuracy when analyzing the single-angle Zimm plots.

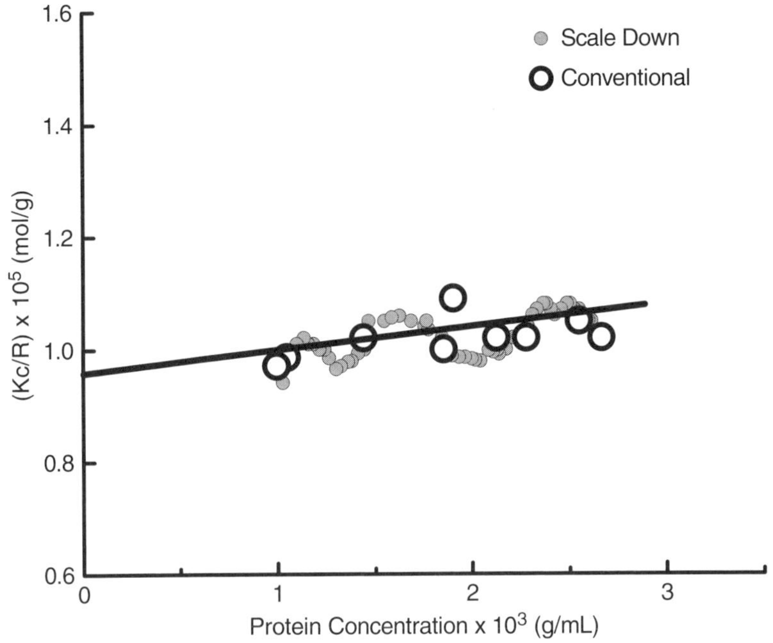

Figure 6.8. Single-angle Zimm plot for Concanavalin A in 0.1 M HEPES buffer as evaluated with both the SD/SLS and conventional SLS techniques.

6.3. Scale Down of Self-Interaction Chromatography

Although the scale down of SLS described above demonstrates a marked improvement in time and sample needed for a single *B* measurement, the method still lacks the intrinsic requirements characteristic of high-throughput analytical methods. For *B* measurements to become a meaningful component of such methods, submicrogram-sample amounts per *B* measurement must be the norm. We now describe the transition to a scale-down version of self-interaction chromatography that meets that requirement.

6.3.1. Evolution of Self-Interaction Chromatography (SIC)

Affinity chromatography, developed in the 1960s, has been loosely defined as a liquid chromatographic technique that utilizes the selective interactions occurring between a binding agent immobilized on the stationary phase (the affinity ligand) and a complimentary mobile phase target (the affinity ligate) to achieve separations.[25] Bioaffinity chromatography exploits the specificity and reversibility of interactions between biological molecules as a means of purification. Examples of systems commonly studied with bioaffinity chromatography include, but are not limited to, the complexes formed by the interactions of enzymes and substrates, antibodies and antigens, and hormones and receptors.

SIC is a specialized adaptation of bioaffinity chromatography that measures homogenous intermolecular interactions using the same protein as the stationary phase and the injected analyte. Chiancone and co-workers are credited with the first report of a SIC-based study.[26] These researchers were able to successfully demonstrate the separation of a mobile-phase mixture of differentially modified hemoglobin using immobilized $\alpha\beta$-hemoglobin-subunit dimers as the stationary phase. Inspired by this pioneering work, Patro et al.[27] applied the principles of SIC to the screening of protein-formulation additives as potential agents for reducing aggregation. This seminal work stressed the technique's sensitivity to weak intermolecular interactions, ease of operation, and high-throughput potential. Shortly after, a review article[28] highlighted SIC as a credible technique for qualitatively assessing protein-protein interactions for purposes of stabilizing and purifying pharmaceutical formulations.

6.3.2. The Synergy of *B* and SIC

The use of SIC to quantitatively measure weak protein-protein interactions in terms of *B* was first reported for lysozyme and chymotrypsin.[29] Later reports characterized the self-interactions for bovine serum albumin (BSA) and myoglobin,[30] determined solution conditions conducive to Ribonuclease A crystal growth,[31] and quantitatively measured interactions between unlike proteins.[32] A key in this effort was the development of a theoretical relationship between SIC chromatographic parameters and *B*, as shown in the following equation.

$$B = \left(\frac{N_A}{MW^2}\right)\left(B_{HS} - \frac{K'}{\phi\rho}\right). \tag{6.2}$$

According to **Eq. 6.2**, *B* can be measured as a function of the molecular weight (MW), Avogadro's number (N_A), protein (hard sphere) excluded volume (B_{HS}), the immobilization density (ρ), the phase ratio (ϕ), and the chromatographic capacity factor (K'). The immobilization density (ρ) is the number of covalently immobilized protein molecules per unit surface area of the bare chromatographic particles. The phase ratio (ϕ) is the surface area of the protein-modified particles that is available to the mobile phase protein. The chromatographic capacity factor (k') is a direct measure of the strength of interaction of the mobile phase analyte with the stationary phase.[33] The equation for K' is given by

$$k' = \frac{V_r - V_0}{V_0}, \tag{6.3}$$

where V_r is the retention volume for the protein species of interest and V_0 is the retention volume for a nonretained species. Unfortunately, an ideal neutral marker has yet to be discovered. Thus, to calculate V_0, the retention volume of acetone in the column containing immobilized protein is multiplied by the ratio of the relative retention volumes of the protein and acetone in a "dead column" made from the same material but with no immobilized protein. This procedure effectively corrects for the difference in size between acetone and protein.

For the case of a globular (i.e., nearly spherical) protein having diameter d, the excluded volume is defined as the volume of solution that is inaccessible to the center of one protein due to the presence of another protein.[34]

$$B_{HS} = 4\upsilon_m = \frac{2}{3}\pi d^3. \tag{6.4}$$

Eq. 6.4 shows that B_{HS} is the equivalent of four molecular volumes of a single protein. Due to the relatively large size of protein molecules, B_{HS} makes a large positive[35] and often dominant[36] contribution to B values.

An initial scale-down version of SIC with PEEK columns[37] has been tested to determine the correlation between SIC and SLS using lysozyme as a model protein system due to the vast extent of literature available for B measurements. A schematic of the general SIC setup is shown in **Fig. 6.9**.

The pump can be either a syringe pump for capillary and microchip systems or a traditional HPLC pump for PEEK columns. The injector is either a standard six-way rotary valve or a nL-injection valve (Upchurch). The SIC active element can be a PEEK, capillary, or microchip column. The detector is a UV absorbance system.

Quantitative agreement between our PEEK/SIC data and previously published SLS data[16,37,38] for lysozyme self-interactions in comparable solution conditions is shown in **Plate 6.1**. Most notable in this work is the time necessary for the experiments. The PEEK/SIC measurements

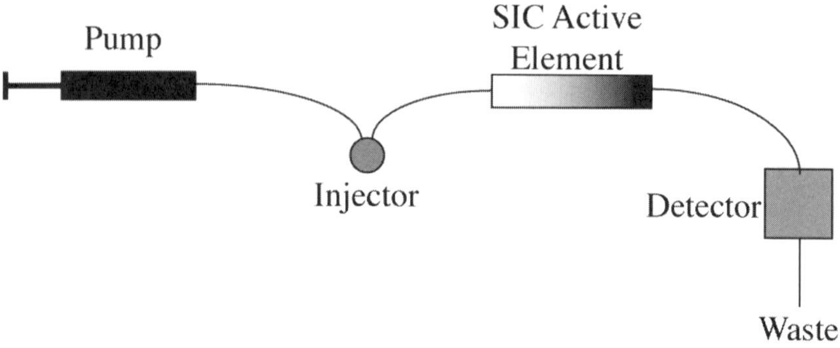

Figure 6.9. Schematic showing basic components of the scale-down SIC experiment.

required less than eight hours for quadruplicate measurements of each of the six conditions tested, i.e., 24 chromatographic runs. The six solution conditions required only two individual solutions (one 0.1 M acetate buffer with 0% NaCl and another with 5% NaCl, both pH 4.5) that were mixed as needed by the HPLC pump. In contrast, one SLS measurement normally requires about two hours (solution preparation plus measurement time) to obtain a single *B* value for a particular solution condition, which extrapolates to about 48 hours in total for quadruplicates of the six conditions. The PEEK/SIC data was reproducible for several weeks, using the same particles, column, and various solvent conditions. The results shown in **Plate 6.1** prove the strong correlation between PEEK/SIC and SLS in the measurement of *B* for lysozyme.

6.4. *B* and Solubility

SIC can make both measurements of *B*—as a function of simple salt solutions and as function of increasingly complex media such as that found in biotechnology, pharmaceutical formulations, and biological systems. Another solution parameter of high interest is temperature, and some proteins have a significant dependence of their solubility on temperature. **Plate 6.2** shows PEEK/SIC results for a lysozyme temperature study. As PEEK/SIC column temperatures are raised there is a clear increase in *B* values. This trend is expected since lysozyme is known to have normal solubility dependence on temperature, i.e., solubility increases with temperature.[12] The correlation between *B* and solubility of proteins is well established.[16,18] While a single *B* measurement cannot be used to predict a corresponding absolute solubility value, the *B* behavior as a function of a particular solution variable mimics the corresponding solubility behavior. For example, as protein-protein interactions become more repulsive (more positive *B* value) there is an increased tendency for proteins to favor protein-solvent interactions leading to higher solubility.

6.5. Potential for Miniaturization

A growing trend in analytical chemistry has been miniaturization, and chromatography has not escaped this trend. In this section, we highlight the current state of the art in miniaturizing SIC methodology. Miniaturization has the potential to significantly reduce protein consumption and analysis time. The first SIC measurements were made using traditional large (5–10 mL volume) glass affinity columns, which required extensive amounts of protein. Three steps have been taken to miniaturize columns for SIC. The first step in scale down was the transfer from a traditional column to one made from common PEEK polymer, as discussed above.[37] The PEEK/SIC columns have a significantly smaller inner diameter (1 mm compared to 4 mm and greater) and consequently a much smaller column volume. This first scale down was an improvement in column volume of more than one order of magnitude while allowing the use of standard HPLC fittings for connection to the remaining instrument. This scale down, however, was insufficient to reduce protein consumption to the microgram scale.

A second approach can be taken to further reduce protein consumption and analysis time by scaling down the PEEK column to a capillary column. **Fig. 6.10** shows a photograph of a CAP/SIC column prepared by the Henry laboratory. This column is made from fused silica and has an inner diameter of 0.25 mm. This further reduction in column volume (80 µL in PEEK to 5 µL in fused silica) allows a 10-fold reduction in the amount of protein immobilized in the column. The smaller columns also allow the injection of smaller sample plugs with the CAP/SIC column routinely using 1 µL injection volumes compared to 10 µL for the PEEK/SIC columns.

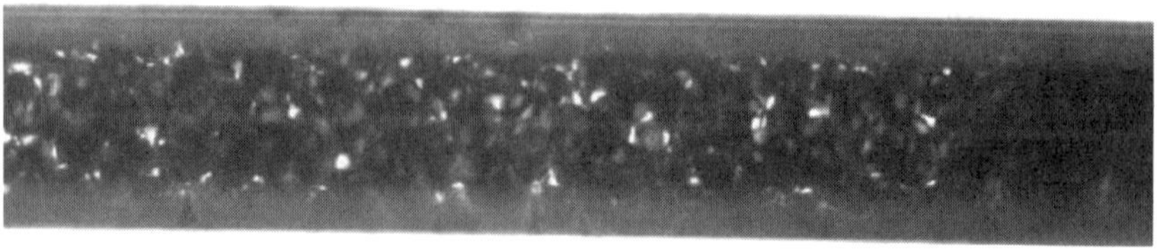

Figure 6.10. Fused silica capillary packed with 65 μm particles coated with lysozyme. Dark area on the right is the polymer frit used to hold the particles in place during runs.

The third scale down for miniaturization is the use of microchip SIC. Microchip columns offer the potential to further reduce the column volume as well as integrate multiple simultaneous analyses in a high-throughput screening approach. Microchip analysis methods in general have been of significant interest in recent years due to the potential to significantly reduce the size and speed of many types analysis.[39] The transition towards microchip SIC has already become a practical avenue for reducing total protein consumption to the microgram scale. **Fig. 6.11** shows a schematic drawing and a photograph of a packed microchip. The use of the microchip allows for the further reduction in protein consumption to less than 1 mg of immobilized protein and less than 40 ng of injected protein per *B* determination. A description of microchip SIC with results regarding *B* measurements has been presented.[40] While not fully developed as a routine method, the microchip method offers great potential for the future of high-throughput, low protein consumption measurements of *B*.

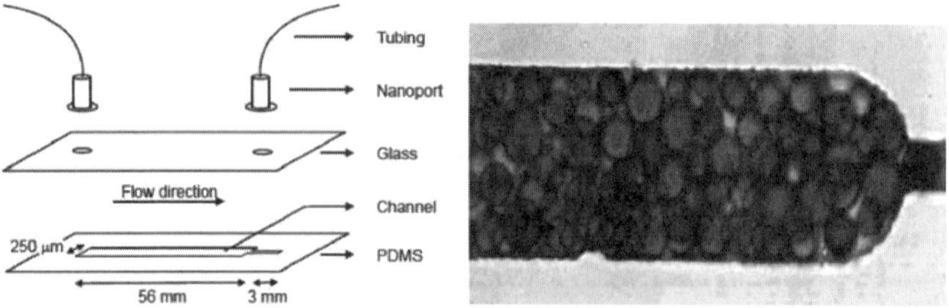

Figure 6.11. (**Left**) Schematic drawing of a SIC microchip. (**Right**) Photograph of a microchip column packed with particles modified with protein.

6.6. Conclusions

The second virial coefficient, B, is a proven and powerful quantitative predictor of protein-protein interactions regarding protein crystallization and protein stability. Current methodologies for measuring *B* are tedious and require inordinate amounts of protein. We have described two scale-down approaches that render *B* measurements more efficient regarding both time and protein requirements. A scale down of static light scattering incorporating simultaneous measurement of scattering intensity and protein concentration allows a single *B* determination in 10–20 minutes using 5–25 μg protein. However, even this scale down of SLS is not compatible with most high-throughput strategies. An alternative approach, scale down of self-interaction chromatography, is suggested that allows the user to define the degree of scale down required for a particular application. PEEK column SIC, glass capillary SIC, and microchip SIC are three progressively scale-down configurations of SIC for which the protein-sample requirement can be as little as 40 ng for a single *B* determination. The SIC configurations, when incorporated with currently available microfluidics, offers the most promising approach for a truly integrated high-throughput method for screening the *B* parameter space.

References

1. Kempner ME, Felder RA. **Proteomics: Diagnostics and specimen biorepositories.** *Am Genomic/Proteomic Tech* 2002, **2:**6–8.

2. Cleland JL, Powell MF, and Shire SJ. **The development of stable protein formulations—A close look at protein aggregation, deamidation and oxidation.** *Crit Rev Ther Drug* 1993, **10:**307–377.

3. Carpenter JF, Kendrick BS, Chang, BS, Manning MC, and Randolph TW. **Inhibition of stress-induced aggregation of protein-therapeutics.** In: ***Amyloids, Prions, and Other Protein Aggregates.*** *Meth Enzymol* 1999, **309:**236–255.

4. Timasheff SN. **Control of protein stability and reactions by weakly interacting co-solvents: The simplicity of the complicated.** *Adv Protein Chem* 1998, **51:**355–432.

5. Chi EY, Krishnan S, Kendrick BS, Chang BS, Carpenter JF, and Randolph TW. **Roles of conformational stability and colloidal stability in the aggregation of recombinant human granulocyte colony-stimulating factor.** *Protein Sci* 2003, **12:**903–913.

6. George A and Wilson W. **Predicting protein crystallization from a dilute solution property.** *Acta Cryst* 1994, **D50:**361–365.

7. Ducruix A, Guilloteau J, Kautt M, and Tardieu A. **Protein interactions as seen by solution X-ray scattering prior to crystallogenesis.** *J Cryst Growth* 1996, **168:**28–39.

8. Malfois M, Bonnete F, Belloni L, and Tardieu A. **A model of attractive interactions to account for fluid–fluid phase separation of protein solutions.** *J Chem Phys* 1996, **105:**3290–3300.

9. Rosenbaum D, Zamora P, and Zukoski C. **Phase behavior of small attractive colloidal particles.** *Phys Rev Lett* 1996, **1:**150–153.

10. George A, Chiang Y, Guo B, Arabshahi A, Cai Z, and Wilson W. **Second virial coefficient as a predictor in protein crystal growth.** *Meth Enzymol* 1997, **276:**100–110.

11. Neal B, Asthagiri D, and Lenhoff A. **Molecular origins of osmotic second virial coefficients of proteins.** *Biophys J* 1998, **75:**2469–2477.

12. Bonnete F, Finet S, and Tardieu A. **Second virial coefficient: Variations with lysozyme crystallization conditions.** *J Cryst Growth* 1999, **196:**403–413.

13. Neal B, Asthagiri D, Velov O, Lenhoff A, and Kaler E. **Why is the osmotic second virial coefficient related to protein crystallization?** *J Cryst Growth* 1999, **196:**377–387.

14. Muschol M and Rosenberger F. **Interactions in under- and supersaturated lysozyme solutions—Static and dynamic light scattering results.** *J Chem Phys* 1995, **103:**10424–10436.

15. Gripon C, Legrand L, Rosenman I, Vidal O, Robert M, and Boue F. **Lysozyme-lysozyme interactions in under- and supersaturated solutions: A simple relation between the second virial coefficients in H_2O and D_2O.** *J Cryst Growth* 1997, **178:**575.

16. Guo B, Kao S, McDonald H, Asanov A, Combs L, and Wilson W. **Correlation of second virial coefficients and solubilities useful in protein crystal growth.** *J Cryst Growth* 1999, **196:**424–433.

17. Demoruelle K, Guo B, Kao S, McDonald H, Nikic D, Holman S, and Wilson W. **Correlation between the osmotic second virial coefficient and solubility for equine serum albumin and ovalbumin.** *Acta Cryst* 2002, **D58:**1544–1548.

18. Haas C, Drenth J, and Wilson W. **Relation between the solubility of proteins in aqueous solutions and the second virial coefficient of the solution.** *J Phys Chem* 1999, **B103:**2808–2811.

19. Ruppert S, Sandler S, and Lenhoff A. **Correlation between the osmotic second virial-coefficient and the solubility of proteins.** *Biotech Progress* 2001, **17:**182–187.

20. Tombs MP and Peacocke AR. ***The Osmotic Pressure of Biological Macromolecules.*** Oxford, UK: Clarendon Press, 1974.

21. Fujita H. ***Foundations of Ultracentrifugal Analysis.*** New York: Wiley, 1975.

22. Kratochvil P. ***Classical Light Scattering from Polymer Solutions.*** Amsterdam: Elsevier, 1987.

23. Wilson WW. **A novel approach for investigating protein solution/crystal equilibrium.** *Eighth International Conference on the Crystallization of Biological Macromolecules*. Florida: Sandestin, 2000.

24. Bajaj H, Sharma V, and Kalonia D. **Determination of second virial coefficient of proteins using a dual-detector cell for simultaneous measurement of scattered light intensity and concentration in SEC-HPLC.** *Biophys J* 2004, **87:**4048–4055.

25. Hage DS. ***Affinity Chromatography. Handbook of HPLC.*** Katz E, ed. New York: M. Dekker, 1998.

26. Chiancone E, Fronticelli C, Gattoni M, Urbaitis B, and Bucci E. **Immobilized hemoglobin in the purification of hemoglobin-based oxygen carriers.** *J Chromatogr* 1992, **604:**117–123.

27. Patro S and Przybycien T. **Self-interaction chromatography: A tool for the study of protein-protein interactions in bioprocessing environments.** *Biotechnol Bioeng* 1996, **52:**193–203.

28. Przybycien TM. **Protein-protein interactions as a means of purification.** *Curr Opin Biotechnol* 1998, **9:**164–170.

29. Tessier P, Lenhoff A, and Sandler S. **Rapid measurement of protein osmotic second virial coefficients by self-interaction chromatography.** *Biophys J* 2002, **82:**1620–1631.

30. Tessier P, Vandrey S, Berger B, Pazhianur R, Sandler S, and Lenhoff A. **Self-interaction chromatography: A novel screening method for rational protein crystallization.** *Acta Crystallogr D* 2002, **58:**1531–1535.

31. Tessier P, Johnson H, Pazhianur R, Berger B, Prentice J, Bahnson B, Sandler S, and Lenhoff A. **Predictive crystallization of ribonuclease A via rapid screening of osmotic second virial coefficients.** *Proteins* 2003, **50:**303–311.

32. Tessier P, Sandler S, and Lenhoff A. **Direct measurement of protein osmotic second virial cross coefficients by cross-interaction chromatography.** *Prot Sci* 2004, **13:**1379–1390.

33. Weston A and Brown P. *HPLC and CE: Principles and Practice*. San Diego, Calif.: Academic Press, 1997.

34. Hermans J. **Excluded-volume theory of polymer protein interactions based on polymer-chain statistics.** *J Chem Phys* 1982, **77:**2193–2203.

35. Neal B and Lenhoff A. **Excluded volume contribution to the osmotic second virial coefficient for proteins.** *AICHE J* 1996, **42:**258–258.

36. Haynes C, Tamura K, Korfer H, Blanch H, and Prausnitz J. **Thermodynamic properties of aqueous alpha-chymotrypsin solutions from membrane osmometry measurements.** *J Phys Chem* 1992, **96:**905–912.

37. Garcia C, Holman S, Henry C, and Wilson W. **Screening of protein-ligand interactions with affinity chromatography.** *Biotech Prog* 2003, **19:**575–579.

38. Rosenbaum D and Zukoski C. **Protein interactions and crystallization.** *J Cryst Growth* 1996, **169:**752–758.

39. Khandurina J and Guttman A. **Microscale separation and analysis.** *Curr Opin Chem Biol* 2003, **7:**595–602.

40. Garcia C, Hadley D, Wilson W, and Henry C. **Measuring protein interactions by microchip self-interaction chromatography.** *Biotech Prog* 2003, **19:**1006–1010.

7

Gearing Optimization Techniques Towards High Throughput

Naomi E. Chayen

Abstract

Considerable advances have been made in methods of protein preparation and of X-ray analysis once diffraction quality crystals are available. These advances have not yet been matched by equally good methods for the crystallization process itself. In the area of crystallization, the main effort and resources are currently being invested into the automation of screening procedures to identify potential crystallization conditions. However, in spite of the ability to generate numerous trials, so far only a small percentage of the proteins produced have led to structure determinations. This is because screening in itself is not usually enough; it has to be complemented by an equally important procedure in crystal production, namely crystal optimization.

In the rush towards Structural Genomics, optimization techniques have been somewhat neglected, mainly because it was mistakenly hoped that large-scale screening would alone produce the desired results. In addition, optimization has relied on particular individual methods that are often difficult to automate and to adapt to high throughput. This chapter addresses a major gap in the field of Structural Genomics by describing practical ways of automating individual optimization methods and adapting them to high-throughput mode.

7

Gearing Optimization Techniques Towards High Throughput

Naomi E. Chayen

Imperial College London, United Kingdom

7.1. Introduction

The last five years have seen some of the greatest achievements in the field of protein crystallization by way of automating and miniaturizing crystallization trials; see[1-5] and **Chapters 2** and **3** in this book. The ability to dispense trials consisting of nanoliter volumes in a high-throughput mode has cut the time of setting up experiments from weeks to minutes, a scenario that was unimaginable just five years ago. Thousands of trials are set up per day and the following phase of image capture and analysis of the crystallization drops is also progressing at high speed (e.g., references[6,7] and **Chapter 9**).

However, in spite of these impressive advances, the crystallization problem has not been solved. High throughput has not led to high output[8] (see **Table 7.1**). Production of suitable crystals is still found to be a rate-limiting step even once proteins have been solubilized and purified. Crystals are often obtained but the problem remains the conversion of those crystals into useful ones. The numbers in the **Table 7.1**, taken from http://targetdb.pdb.org show the sum of results from 17 structural genomics/proteomics projects worldwide on January 3, 2007. This web-

Table 7.1. The steps leading to X-ray structure determination (the numbers are taken from http://targetdb.pdb.org).

Summary of all major Structural Genomics projects 3 January 2007	
Cloned	87039
Expressed	51816
Soluble	23793
Purified	20296
Diffracting crystals	3808
Structure defined	2814

From **http://targetdb.pdb.org**

site provides timely status and tracking information on the progress of the production pipeline and the solution of structures. (The numbers for each individual project vary as they are dealing with different genomes and have started at different times). The figures demonstrate that out of 87,039 clones and 51,816 expressed proteins, 23,793 proteins were soluble and 20,296 were purified. Of these, 3,808 good crystals were produced leading to the determination of 2,814 structures so far. The website gives a more detailed picture of the crystallization showing that 7,713 of the trials that were set up from pure soluble proteins had produced crystals, yet only 3,808 of those crystals were useful ones.

7.2. Development of Optimization Procedures

Once a "lead" is obtained from screening trials, optimization can be performed in one of two ways:[8]

(a) The most common way is to fine-tune the crystallization conditions by varying the parameters involved such as the concentrations of protein, precipitants, pH, temperature, etc. This in fact is also screening that focuses on a more defined range of conditions.

(b) A second means of optimization is to actively influence and control the crystallization environment while the trial takes place in order to lead crystal growth in the direction that will give the best results.

Screening procedures can readily be automated and adapted to high throughput as detailed in the **Chapters 2** to **4**. These procedures are very valuable as they are essential in order to find initial conditions for crystallization and in some cases can also produce high-quality crystals. However, when screening and subsequent optimizing by fine-tuning of the initial screening conditions fails, it is necessary to implement the second approach—that of optimizing the crystallization conditions by actively controlling the crystallization process.[9,10] Such methods are currently conducted manually as they are more complicated and do not easily lend themselves to automation. There is however an urgent need to find ways to automate and miniaturize these techniques in order to cope with the vast numbers of "leads" resulting from the screening procedures. To date only very few such optimization methods have been automated and some of these are described below.

7.2.1. Crystallization in Gels

Growth of crystals in a gel medium can improve the quality of crystals in comparison with solution media because convection and sedimentation are reduced[11] (see also **Chapter 5**). The application of gels in the crystallization of macromolecules, particularly proteins, has been pursued for more than ten years, yet the method remains underused. This may be due to the relatively complicated procedures required when applying gels to crystallization trials and also to the large quantities (mostly greater than 10 μl) of sample needed. The last two years have seen major improvements in the use of gels by miniaturization as well as automation of gel trials. Crystals can be grown in small volumes of gel inside capillaries, thus combining the advantages of growth in gel with those of the FID method[12] (see also **Chapter 5**). A simpler means, described here is to set up gelled trials in microbatch, which has enabled for the first time, the automatic dispensing of 0.3–2 μL gelled drops in high-throughput mode.[13]

Two advances are described here: (i) Automation of crystallization in gels (ii) Generation of gelled drops using very small volumes (≥ 0.3 μL). These have been achieved using the IMPAX robot.

IMPAX[14] works as a four- or five-channel system where precipitant, buffer, protein, etc. are put into different channels and dispensed through a fine multi-bore tip by the action of motorized syringes. By placing a gel solution in one of the channels it is possible to automatically dispense microbatch trials, which form the gelled crystallization mixtures in final volumes of 0.3–2 μL. This can be achieved with the same ease as conventional automated microbatch trials. The gel solution is loaded into the liquid handling apparatus in the same way as the other ingredients of the crystallization trial. Then, while still a low-viscosity liquid, it is dispensed under oil simultaneously with all the other ingredients. After a known time polymerization occurs and the drop gels.[15] Agarose gels and two silica gels, one made by neutralization of sodium metasilicate and the other by hydrolysis of tetramethyl ortho silane (TMOS), have been tested with trypsin, thaumatin, concanavalin A, lysozyme, and C-phycocyanin (see **Table 7.2**).[13]

The results show that a higher proportion of the total crystal yield consisted of large usable crystals in the gelled drops with at least one of the gels (**Table 7.2**). Trials were dispensed in volumes of 0.3–3 μL. Best results were observed between 0.7 and 3 μL (see **Plate 7.1**). In order to test whether the gels would make a difference even under sub-optimal

Table 7.2. Crystals in standard and gelled microbatch drops (from ref.[13] with permission).

Protein	Most effective types of gel	Conditions used for crystallisation	Protein conc. (mg/mL)	Typical crystal numbers and sizes(μm), in microbatch	
				gelled	standard
C-phycocyanin 126 kDa	TMOS 0.2%(v/v)	0.7–1.0 M ammonium sulphate 40 mM MES, pH 6.1 1.5 mM dodecyl maltoside	10–20	40 140×140×110	150 140×100×80 (often smudged)
Thaumatin 22 kDa	TMOS 0.2%(v/v) metasilicate 0.2%(v/v)	5–13%(w/v) Na/K tartrate 50 mM PIPES, pH 6.8	20–25	10 500×300×250	25 400×200×150
Trypsin 24 kDa	TMOS 0.2%(v/v) agarose 0.1%(w/v)	33–36%(sat) ammonium sulphate 100 mM Tris, pH 8.5	20–35	1 700×200×150	5 (frequent clusters) 700×200×100 (generally flat)
Lysozyme 14.5 kDa	TMOS 0.2%(v/v)	4.6–7%(sat) sodium chloride 40 mM sodium citrate, pH 4.6	20	1 400×300×300	5 400×300×300
Concanavalin A 102.5 kDa	TMOS 0.2%(v/v)	(a) 3–10%(w/v) PEG 6K, 100 mM Na cacodylate, pH 5.6 (b) 52–60%(sat) ammonium sulphate 100 mM phosphate buffer, pH 7	(a) 5–10 (b) 10	50 160×40×40	150 80×20×20

(e.g., initial screening) conditions, concanavalin A was crystallized in the presence of either agarose or TMOS gels at conditions that normally gave many small crystals. The presence of TMOS gel made a striking difference; fewer, larger crystals were obtained (see **Table 7.2** and **Plate 7.1e, f**). TMOS at a concentration of 0.2% has consistently given the best results with all five proteins tested.[13]

7.2.2. "Containerless" Crystallization

Heterogeneous nucleation, which is often detrimental to the production of diffraction quality crystals, can be induced by the contact of a crystallization trial with the walls of its supporting vessel.[16] Crystallization in a "containerless" set up in which a crystallization drop is suspended between two oils of different densities results in reduction of heterogeneous nucleation[17,18] thus leading to the production of a smaller number of high-quality crystals. This two-oil method was useful for manually setting up a small number of experiments, but it took time to layer the oils, it was difficult to harvest the drops, and the method was not amenable to automation.[13]

High throughput can now be achieved by a modification of the original method whereby the lower layer, previously consisting of a heavy fluorosilicone oil, is replaced with a much cheaper hydrophobic surface of vaseline. The vaseline covers the bottom of the crystallization plate onto which the trial drops (0.3–2 μL) are automatically dispensed using a robot. MicroWell Modules (Nunc Intermed, Denmark) and Linbro plates have been used, but most other crystallization plates are suitable (**Plate 7.2**). A low-density oil, usually paraffin (0.84 g/cm^3), is applied as the top layer. The greased surface has an added advantage over the orignal two-oil set up, in that the grease provides a stable interface to the upper layer. This prevents crystals from migrating to the walls, making them much easier to harvest. The crystals could be lifted directly out of the drop with a loop or even a spatula. This method was tested for several proteins including a newly crystallized C-phycocyanin. In the "containerless" set up, prisms of C-phycocyanin measuring 0.1 × 0.1 × 0.1 mm^3 were obtained at crystallization conditions, which yielded mostly thin "un-X-rayable" plates in standard microbatch drops. In some cases, the drops in contact with the vessels yielded crystals measuring only 0.06 mm in the largest dimension.[13]

It was also noticed that many drops could be dispensed within a small flat area coated with the grease, doing away with the need for wells (e.g., ten drops were dispensed on a 0.95 cm^2 flat area).

7.2.3. Control of Evaporation Kinetics

It is well known that nucleation is a prerequisite for and the first step in crystal growth,[19-21] yet excess nucleation yields a large number of small crystals instead of a small number of useful ones. A means of controlling nucleation by reaching nucleation slowly and then stopping it before it becomes excessive can now be carried out in microbatch. This is achieved by controlled evaporation, and therefore concentration, of the drops through a thin oil layer. Evaporation is later arrested by increasing the thickness of the oil layer.

The paraffin oil generally used in standard microbatch trials is not completely impermeable to the aqueous solution that constitutes the crystallization drops. The conventional microbatch method involves using a layer of oil thick enough (4 mm, corresponding to 8 mL covering all 72 wells of a microbatch plate measuring $8 \times 5.5 \times 0.9$ cm^3) to render evaporation through it negligible within the time scale of a crystallization experiment (typically one week to one month). However, if controlled evaporation is required, the thickness of the layer can become an active parameter of the process. Instead of setting the microbatch conditions well inside the nucleation zone of the phase diagram, conditions are set to be undersaturated or metastable and water is allowed to evaporate slowly through a thin oil layer (1.2–2 mm for 2-μL drops, **Plate 7.3a**). This optimum range for the oil layer thickness has been determined by extensive trials and seems to be independent of the protein used. It is of course a function of the volume of the original drop. The solution therefore arrives at the nucleation zone in a controlled way. The thickness of the oil layer is then increased, rendering evaporation negligible, and the experiment progresses along the conventional batch route (**Plate 7.3b**). Assuming that the evaporation has been arrested at the early stages of nucleation, this means that the trial spends most of its lifetime in the metastable zone of conditions, as the protein gets absorbed into the forming crystals. This procedure can be conducted automatically with an extra arm added to the crystallization robot for dispensing oil after set time intervals.[13]

So far this procedure has been tried with trypsin, lysozyme, and C-phycocyanin. The metastable starting conditions are listed in **Appendix 7.1** "Experimental Procedures" The critical times after setup at which to arrest evaporation by adding more oil were found to be 18–24 h for trypsin, 2–4 h for lysozyme and 4–6 h for C-phycocyanin (for 2-μL drops). Trials set at the same conditions without allowing any evaporation resulted in clear drops. If trials were allowed to evaporate without arresting, showers and eventually drying-out of the drops occurred. Arresting the evaporation at the above times enhanced the size and yield of useful crystals compared with crystals grown in the standard microbatch method.[13]

7.2.4. Optimization by Decoupling Nucleation and Growth

Nucleation requires different conditions to those of growth. The most common way of decoupling nucleation and growth is by seeding.[22] However, quenching of nucleation using dilution is more amenable to high-throughput processing. Dilution can be achieved in both microbatch[23] and hanging drop methods.[24] In the case of hanging drops, the coverslips holding the drops are incubated for some time over reservoir solutions that normally give many small crystals. After a given time the coverslips are transferred over reservoirs with lower precipitant concentrations that would normally yield clear drops. Various studies have shown that there is a time lag between the solution reaching the appropriate supersaturation and the formation of the first postcritical nuclei.[25] This induction time for nucleation is however fairly short compared to the time at which the crystals start becoming visible in a microscope[23]. In practice, for the vapor diffusion experiments, one must allow sufficient evaporation against the high-concentration reservoir to take place so that the drop reaches spontaneous nucleation levels, and add to that a little more time to allow for the nucleation time lag. To avoid shock to the drop due to the transfer, the low-concentration reservoir solutions are dispensed at the beginning of the experiment and sealed with self-adhesive tape. This allows a humid atmosphere to develop above the well to which the coverslip will be transferred. The transfer itself lasts 1–2 seconds. In the case of microbatch, the drops are diluted by automated means after incubation. Successful use of these techniques have so far been reported only in highly individualized experiments where nucleation and metastable conditions, as well as optimal timing, had been fine-tuned to the system in hand.[23,26,27] The suitability of this approach has now been tested for high-throughput trials. Hanging drops were transferred at various times (selected by reference to the time, which it took to see the first crystals in the preliminary screens) from a standardized set of screening solutions at high concentrations to screens at lower concentrations (see "Experimental Procedures"). Trials using model proteins indicate that the average number of trials leading to good crystals is higher when using this method, compared with using a screen consisting of only either the high- or the low-concentration sets of conditions. For a decoupling experiment to be successful, it must be ensured that the solution is diluted to the metastable zone of conditions after incubation at spontaneous nucleation conditions. It was however found that when these conditions are not known beforehand, as is the case in a high-throughput environment, the best results were obtained when standard screening kits solutions were diluted to between 60 and 80%. These observations have led to the development and use of a "3D screen," which has an inbuilt first optimization step (see "Experimental Procedures"). These experiments can now be performed automatically with robots for setting up hanging drops (**Plate 7.4**).[13]

7.2.5. Slowing Down Vapor Diffusion with an Oil Barrier

A means to slow down the equilibration rate and thereby approach supersaturation more slowly in order to avoid crystal "showers" is by placing a paraffin/silicone oil mixture as a barrier over the reservoir of a hanging or sitting drop trial.[28] The mix of paraffin and silicone oil can be varied as needed. It was found that volumes of 250–500 μL placed over 1-mL reservoirs in standard Linbro plates (corresponding to layer thickness of 1.2–2.5 mm) were most efficient. This method has been shown to work well for several proteins.[29,30] The advantage of this technique is that no change is required neither to the crystallization conditions nor to the method used. It can be applied in Linbro, VDX, Cryschem, or any other vessel, and it can easily be automated by adding one extra step to the procedure used by existing crystallization robots.[13]

7.3. Summary

Numerous recent books, articles, and special issues of scientific journals have highlighted the importance of Structural Genomics in the post-Genome era. However, surprisingly little attention has been given to improving methods of protein crystallization and optimization. The statistics shown in **Table 7.1** and the colossal number of proteins that still need to be dealt with, indicate that, like screening, optimization must be adapted to high throughput. Without this we will soon run out of resources and be left with a backlog of useless microcrystals. The combination of automated screening with further development of automated crystal optimization methods will remove the main bottleneck in Structural Genomics and equip the Genome Projects to move forward into the post-genomic era.

References

1. Bodenstaff ER, Hoedemaeker FJ, Kuil ME, de Vrind HP, and Abrahams JP. **The prospects of protein nanocrystallography.** *Acta Crystallogr D Biol Crystallogr* 2002, **58**(Pt 11):1901–1906.

2. Walter TS, Diprose JM, Mayo CJ, Siebold C, Pickford MG, Carter L, Sutton GC, Berrow NS, Brown J, Berry IM et al. **A procedure for setting up high-throughput nanolitre crystallization experiments. Crystallization workflow for initial screening, automated storage, imaging and optimization.** *Acta Crystallogr D Biol Crystallogr* 2005, **61**:651–657.

3. Hansen CL, Skordalakes E, Berger JM, and Quake SR. **A robust and scalable microfluidic metering method that allows protein crystal growth by free interface diffusion**. *Proc Natl Acad Sci USA* 2002, **99**(26):16531–16536.

4. Stevens RC. **High-throughput protein crystallization**. *Curr Opin Struct Biol* 2000, **10**(5):558–563.

5. Santarsiero BD, Yegian DT, Lee CC, Spraggon G, Gu J, Scheibe D, Uber DC, Cornell EW, Nordmeyer RA, Kolbe WF et al. **An approach to rapid protein crystallization using nanodroplets**. *J Appl Crystallogr* 2002, **35**:278–281.

6. Luft JR, Collins RJ, Fehrman NA, Lauricella AM, Veatch CK, and DeTitta GT. **A deliberate approach to screening for initial crystallization conditions of biological macromolecules**. *J Struct Biol* 2003, **142**(1):170–179.

7. Jurisica I, Rogers P, Glasgow JI, Collins RJ, Wolfley JR, Luft JR, and DeTitta GT. **Improving objectivity and scalability in protein crystallization.** *IEEE Intelligent Systems* 2001, **16**(6):26–34.

8. Chayen NE. **Protein crystallization for genomics: Throughput versus output**. *J Struct Funct Genomics* 2003, **4**(2-3):115–120.

9. Chayen NE. **Turning protein crystallisation from an art into a science**. *Curr Opin Struct Biol* 2004, **14**(5):577–583.

10. Chayen NE. **Methods for separating nucleation and growth in protein crystallisation**. *Prog Biophys Mol Bio* 2005, **88**(3):329–337.

11. Robert MC, Vidal O, Garcia-Ruiz JM, and Otlalara F. ***Crystallisation in gels and related methods***. Oxford, United Kingdom: Oxford University Press, 1999.

12. Ng JD, Gavira JA, and Garcia-Ruiz JM. **Protein crystallization by capillary counterdiffusion for applied crystallographic structure determination**. *J Struct Bio*l 2003, **142**(1):218–231.

13. Chayen NE and Saridakis E. **Protein crystallization for genomics: Towards high-throughput optimization techniques**. *Acta Crystallogr D Biol Crystallogr* 2002, **58**(Pt 6 Pt 2):921–927.

14. Chayen NE, Stewart PDS, Maeder DL, and Blow DM: **An automated-system for microbatch protein crystallization and screening**. *J Appl Crystallogr* 1990, **23**:297–302.

15. Moreno A, Saridakis E, and Chayen NE. **Combination of oils and gels for enhancing the growth of protein crystals**. *J Appl Crystallogr* 2002, **35**:140–142.

16. Yonath A, Mussig J, and Wittmann HG. **Parameters for crystal growth of ribosomal subunits.** *J Cell Biochem* 1982, **19**(2):145–155.

17. Chayen NE. **A novel technique for containerless protein crystallization**. *Protein Eng* 1996, **9**(10):927–929.

18. Lorber B and Giege R. **Containerless protein crystallization in floating drops: Application to crystal growth monitoring under reduced nucleation conditions.** *J Cryst Growth* 1996, **168**:204–215.

19. McPherson A. ***Crystallization of Biological Macromolecules.*** Cold Spring Harbor, New York: Cold Spring Harbor Laboratory Press, 1999.

20. Ducruix A and Giegâe R. ***Crystallization of Nucleic Acids and Proteins: A Practical Approach***, 2nd ed. Oxford, United Kingdom: Oxford University Press, 1999.

21. Bergfors TM. ***Protein Crystallization: Techniques, Strategies, and Tips. A Laboratory Manual***. La Jolla, California: International University Line, 2001.

22. Stura E. **Seeding techniques.** In: ***Crystallization of Nucleic Acids and Proteins: A Practical Approach***, Ducruix A and Giegâe R, eds, 2nd ed. Oxford, United Kingdom: Oxford University Press, 1999.

23. Saridakis EE, Stewart PD, Lloyd LF, and Blow DM. **Phase diagram and dilution experiments in the crystallization of carboxypeptidase G2.** *Acta Crystallogr D Biol Crystallogr* 1994, **50**(Pt 3):293–297.

24. Saridakis E and Chayen NE. **Systematic improvement of protein crystals by determining the supersolubility curves of phase diagrams.** *Biophys J* 2003, **84**(2 Pt 1):1218–1222.

25. Rosenberger F, Howard SB, Sowers JW, and Nyce TA. **Temperature-dependence of protein solubility—Determination and application to crystallization in X-ray capillaries.** *J Cryst Growth* 1993, **129**(1–2):1–12.

26. Saridakis E and Chayen NE. **Improving protein crystal quality by decoupling nucleation and growth in vapor diffusion.** *Protein Sci* 2000, **9**(4):755–757.

27. Nield J, Rizkallah PJ, Barber J, and Chayen NE. The **1.45 Å three-dimensional structure of C-phycocyanin from the thermophilic cyanobacterium *Synechococcus elongatus*.** *J Struct Biol* 2003, **141**(2):149–155.

28. Chayen NE. **A novel technique to control the rate of vapor diffusion, giving larger protein crystals.** *J Appl Cryst* 1997, **30**:87–91.

29. Mandelman D, Gonzalo P, Lavergne JP, Corbier C, Reboud JP, and Haser R. **Crystallization and preliminary X-ray study of an N-terminal fragment of rat liver ribosomal P2 protein.** *Acta Crystallogr D Biol Crystallogr* 2002, **58**(Pt 4):668–671.

30. Isupov MN, Brindley AA, Hollingsworth EJ, Murshudov GN, Vagin AA, and Littlechild JA. **Crystallization and preliminary X-ray diffraction studies of a fungal hydrolase from *Ophiostoma novo-ulmi*.** *Acta Crystallogr D Biol Crystallogr* 2004, **60**(10):1879–1882.

Appendix 7.1.

Experimental Procedures

Porcine pancreatic trypsin (T-0134), thaumatin from *Thaumatococcus danielii* (T-7638), Jack Bean type IV concanavalin A (C-2010), hen egg-white lysozyme (L-6876), buffers, and salts were purchased from Sigma. The C-phycocyanin, from *Synechococcus elongatus*, was purified in house (J. Nield, personal communication). All experiments were performed at 18°C.

A7.1.1. Preparation of Gels[13]

2-mL stock solutions of gels were prepared to load the robot for each series of experiments:

(i) Agarose 1% (w/v) stock: 20 mg "Wide Range" agarose powder (Sigma, cat. no. A2790) was progressively dissolved into 2 mL deionized water previously heated to approximately 80°C and continuously stirred. The optimal concentration of agarose in the crystallization drop was 0.1–0.15% w/v.[15] Higher concentrations often caused precipitation.

(ii) TMOS 5% (v/v) stock: 0.1 mL TMOS solution (Fluka, cat. no. 87682) was added to 1 mL distilled water in a glass tube, and the solution vigorously stirred to disperse. It was then topped up to 2 mL and vigorously stirred for an additional 10–15 min, kee-ping the vessel covered. The optimum final concentration in the crystallization drop was found to be 0.2% v/v.[15]

(iii) Sodium metasilicate 5% (v/v) stock: silicate solution (Sigma, cat. no. 33844-3), which has an initial pH of 11.6, was diluted with distilled water at almost the desired concentration and acidified to pH 6.5 by addition of 1 M acetic acid solution while stirring. The optimum final concentration in the drop was 0.2% as for TMOS (A. Moreno, Personal communication).

A7.1.2. Control of Evaporation[13]

0.3-2-μL microbatch drops were set up at the following conditions (where the pH indicates pH of the buffer and not of the final mixture, and the saturation temperature of the ammonium sulfate solution was 20°C).

Lysozyme: 20 mg/mL protein, 50 mM sodium acetate-HCl pH 4.6; 0.7 M sodium chloride.

C-phycocyanin: 15 mg/mL protein, 40 mM MES-NaOH buffer pH 6.1; 1.5 mM dodecyl maltoside, 0.6–0.72 M ammonium sulfate.

Trypsin: 20 mg/mL protein, 50 mM Tris-HCl buffer pH 8.4; 39% (sat) ammonium sulfate.

A7.1.3. Decoupling of Nucleation and Growth[13]

The 3D Screen (Molecular Dimensions, United Kingdom, MD1-13) is made of 48 solutions. 24 contain sparse matrix screening conditions; the other 24 are a 70% dilution of the precipitants in those (buffer and additive concentrations are kept constant). Using this screen, all hanging drops were first incubated for 3–6 h over the solutions at high concentrations. The coverslips holding the drops were then transferred over the reservoirs at 70% dilution.

8

Miniaturization and Automation for High-Throughput Membrane Protein Crystallization in Lipidic Mesophases

Vadim Cherezov and Martin Caffrey

Abstract

A robot for setting up crystallization trials with an emphasis on membrane proteins is described. The robot was assembled from commercially available components and is relatively inexpensive. It was designed to perform crystallizations of membrane proteins from within the highly viscous lipidic cubic mesophase (*in-meso* or cubic-phase method) and it does so in a way that is extremely economical in terms of protein, lipid, and precipitant. Standard crystallization conditions call for just 20 nL protein solution, 30 nL lipid, and 1 μL precipitant solution. Where necessary, it is possible to reduce lipid and protein volumes by a factor of 2.5. Under standard conditions, precipitant- and mesophase-volume delivery is reproducible.

The crystallization robot is quite versatile in that it can be used not only for *in-meso* trials but also for vapor diffusion, microbatch, and bicelle crystallizations. Further, the robot works equally well with membrane and soluble proteins. The robot is also used to prepare screen solutions.

96-well glass-based crystallization plates have been designed and built for use with the *in-meso* robot. They have exquisite optical properties that lend themselves to visualizing microcrystals and to birefringence-free examination between crossed polarizers. The plates have a very small profile height facilitating high-density storage. The plate footprint and 96-well layout follow SBS standards. The polyester tape used to create the wells is watertight. Thus, dehydration of well contents is not a problem when storage is carried out under moderate conditions. Well contents are sandwiched between two glass plates and are mechanically quite stable such that the plates do not require careful handling. The glass plates are themselves reusable.

To cope with the high throughput of trials made possible by the crystallization robot, an automated imaging system has been designed and built, once again using commercially available components. Its optics are matched to the well thickness of the glass crystallization plates such that the complete volume of the mesophase bolus can be imaged in a single shot. A 96-well plate can be imaged in 4.5 min. The slower image processing is done subsequently off-line.

The robots described here enable high-throughput crystallization trials of membrane proteins using lipidic mesophases with unprecedented speed and with microgram quantities of protein. This means that more membrane protein targets will be subjected to crystallization trials by way of the *in-meso* method with relatively fast turnaround. As more targets are examined the utility of the *in-meso* method for membrane protein crystallization will be more completely evaluated. This automation will contribute to broadening the range of membrane proteins that yield to structure determination and is in line with the NIH Structural Biology Road Map (see **Chapter 1** in this book). Success in obtaining crystals, and ultimately the atomic structure of membrane proteins, will enhance our understanding of some of the most fundamental processes underlying cellular function that are integral to living systems.

8

Miniaturization and Automation for High-Throughput Membrane Protein Crystallization in Lipidic Mesophases

Vadim Cherezov[*] and Martin Caffrey[†]

[*] ***The Scripps Research Institute, La Jolla, California***
[†] ***University of Limerick, Limerick, Ireland***

8.1. Introduction

Crystallography is the most reliable way to explore membrane protein structure–function relationships in atomic detail. To do crystallography requires crystals, which are not readily procured, particularly in the case of membrane proteins. Part of the problem is the dearth of usable quantities of pure and active membrane proteins. The other concerns the production of diffraction quality crystals.

To date, nearly 100 different membrane protein structures have been posted in the Protein Data Bank (www.rcsb.org). The bulk of these have been determined with crystals grown using well-established vapor-diffusion techniques in combination with an assortment of solubilizing surfactants (herein referred to as the *in-surfo* method). Over the past several years, newer methods have emerged that exploit the ability of lipids to form liquid crystals or mesophases and to reconstitute membrane proteins (reviewed in the article[1]). These are collectively referred to as the *in-meso* methods. One of these is based on the lipidic cubic phase[2] while another makes use of lipid/detergent bicelles.[3] The former has produced crystal structures of a number of important membrane proteins and helped elucidate the photocycle in the light-driven proton pump, bacteriorhodopsin, in atomic detail. While well known for its ability to produce crystals of proteins consisting of transmembranal helices, the method has only recently been shown to produce diffraction-quality crystals of the vitamin B_{12} transporter, BtuB, a bacterial outer membrane β-barrel protein.[4] The highest-resolution structure reported for this modification of the *in-meso* method is that of the K intermediate of bacteriorhodopsin (bR; 1.43 Å).[5] Other membrane proteins have produced crystals or microcrystals by the *in-meso* method that diffract poorly or do not diffract at all. These include the acetylcholine-bungarotoxin complex[6] and the reaction center from *Blastochloris* (formerly *Rhodopseudomonas*) *viridis*.[7] To put the method in perspective, of a total of 463 integral membrane protein entries in the PDB, 48 are attributed to the *in-meso* approach (www.mpdb.ul.ie).[8]

Despite the successes, the yield of the method has not been impressive. The sense is that this, in part, relates to the perceived difficulty in performing *in-meso* crystallization and the reluctance on the part of the community members to use the method. Much of the hesitancy originates from the fact that the method involves manipulating extremely viscous lipidic dispersions. At another level, reluctance to employ the method likely originates from a lack of an understanding of the underlying mechanism whereby crystallization occurs *in meso*. An hypothesis for how this happens at the molecular level has been advanced[1] (**Plate 8.1**). The robot described in this chapter is facilitating the testing of the proposal.

Recently, we have introduced a mixing device that facilitates the handling of small quantities of such viscous lipidic materials.[9] This has been adapted to *in-meso* crystallization. Moreover, we developed inexpensive, glass-based crystallization plates for use with the lipid mixing and a dispensing device that enables *in-meso* crystallization trials of membrane proteins to be set up with ease.[10] The latter advances however are limited by virtue of the fact that the entire process must be done by hand. Thus, manual dexterity, patience, and good visual acuity are required of the user, especially when nanoliter volumes of materials are being dispensed. For many membrane proteins, crystallizations must be set up under "hostile" conditions such as at frigid temperatures and in the dark or under reduced light. Throughput is therefore severely limited by the current instrumentation.

Crystallization trials are characterized by the need to process large numbers of samples as the multidimensional crystallization space is explored for crystal nucleation and growth. With the *in-meso* methods there is the additional need to include lipid identity and concentration as a variable to be explored. This only adds to the number of trials that must be performed. There is also pressure to miniaturize and to scale down when membrane proteins and lipids are involved since they come at a premium.

We have responded to these assorted challenges—the need for automation, high throughput, and miniaturization—by building a robot for *in-meso* crystallization of membrane proteins.[11] The robot was designed to meet the above specifications and for a limited budget. It consists for the most part of commercially available components that have been assembled in a way that realizes these objectives. The robot is robust and performs reliably. Herein, we report on our experience with the equipment over a one-year period. Home-made, glass-based 96-well plates have been built for use with the robot. They offer unparalleled optical performance and allow for the detection of microcrystals and for birefringence-free viewing. With a profile height of less than 2 mm, high-density plate storage is possible. Under standard conditions, *in-meso* crystallization trials are performed with 50-nL protein/lipid mesosphase [representing 20 nL protein solution (200 ng protein at 10 mg protein/mL)] and 1-μL precipitant solution per trial. A single 96-well plate can be set up in 13 min with a 4-tip robot and in 8 min with an 8-tip machine. The robot is quite versatile and has been adapted to perform vapor diffusion, microbatch, and bicelle crystallizations of membrane and soluble proteins. It is also being used to prepare solutions for screening, as described below (**Section 8.2.2**).

In addition to the crystallization robot, we have built an imaging system for automatic recording and processing of well-content images. It has an optical system with a depth of field that matches sample thickness so that only a single image must be recorded to view the entire well contents. Image acquisition from a 96-well plate takes less than 5 min. Both the crystallization and the imaging robot are described in **Section 8.2**

8.2. The *In-Meso* Crystallogenesis Robot Suite

8.2.1. Crystallization Robot

The robot was designed and built first and foremost to perform automated crystallization of membrane proteins using the *in-meso* method. However, as has been demonstrated elsewhere,[11] it does work with other crystallization protocols. The robot was built based on a commercially available liquid handling robot (**Plate 8.2**). The open style, modular design, and software flexibility allowed us to adapt it so that it would perform the steps needed for *in-meso* crystallization. The standard robot consists of a 1 × 0.7 m deck with two independent arms. Arm 1 is used for liquid aspiration and dispensing while Arm 2 is typically a plate gripper and moving device. We retained the original function of Arm 1 and equipped it with four small-volume liquid handling tips for pickup and placement of precipitant solutions. Arm 2 was modified to handle the viscous protein/lipid mesophase. To this end, the gripper device was replaced with a motor-driven positive-displacement Hamilton syringe. The controller for the latter was interfaced to the liquid dispensing robot through an RS-232 cable. The positional accuracy of Arm 2 was raised by increasing the gear ratio of the *z*-motor by a factor of eight.

A second robot has been assembled modeled on the one just described. Instead of 4, it was built to include eight small-volume liquid handling tips. Its performance characteristics are described below (**Section 8.2.2**).

8.2.2. Screen Solution Preparation

The liquid handling features of the crystallization robot introduced above (see **Section 8.2.1**) has been adapted for use in customized screen solution preparation. Stock solutions can be housed in 5-mL tubes or in 96-deep-well blocks, and screens are prepared typically in 100-μL capacity 96-well plates. Programs have been written for grid and random screen formulation. In the former, up to 24 individual screen components can be used and included in the final screen at defined concentrations. In the random screen mode, up to 96 different stock solutions can be used. Components are mixed randomly according to user-defined rules. These include concentration range, frequency of occurrence of each component, and forbidden combinations. Typically, 96 screen solutions can be prepared in less than 20 min.

Considerable care was taken to ensure volumetric accuracy in screen preparation. To this end, all pipettes were calibrated gravimetrically for aspiration and delivery with 1, 10, and 100 μL of water. Subsequently, liquids with varying viscosity characteristics were investigated to arrive at a set of liquid handling parameters that would provide accurate delivery of the broadest range of typical precipitant solutions. However, we were not able to handle satisfactorily large (greater than 10 μL) volumes of high-viscosity liquids, such as greater than 25% PEG 4000 and greater than 30% MPD. The problem arose because the current *in-meso* application calls for small bore diameter pipettes to ensure accurate delivery of small (equal or less than 1 μL) precipitant solution volumes. Where really viscous materials must be included in the screen solutions, the positive displacement features of the syringe on Arm 2 of the robot will be implemented.

8.2.3. Crystallization Plates

The plates were designed specifically for use with the *in-meso* robot and were modeled on a similar, but smaller, plate developed for manual *in*-meso crystallization.[10] The plate has three components: a flat glass base plate, a perforated 96-hole polymeric spacer, and a glass coverslip (**Plate 8.3a**). The base plate has a footprint of 127.8 × 85.5 mm and conforms to the Society for Biological Screening (SBS) standard for microplates. The plate is 1 mm thick. The spacer is made from double-stick (acrylic glue) polyester film that is 0.13 mm thick. The film is cut into sections 112 × 77 mm and contains 96 holes each 5 mm in diameter with a standard 9-mm separation (**Plate 8.3a**). The spacer sits atop the base plate and, when filling is complete, the glass coverslip (0.2 mm thick) is placed over and onto the spacer creating 96 individually sealed and isolated wells. The cubic phase bolus is sandwiched between and in intimate contact with the upper and lower glass plates. This arrangement gives the best possible optical properties that allow to observe well contents for crystal growth. An example of a loaded and sealed plate set up for *in-meso* crystallization is shown in **Plate 8.3b**.

8.2.4. Imaging Station

The imaging station was designed in-house. It includes the following components (**Plate 8.4**): a 12× zoom lens, a 3.3 MP cooled CCD camera, an *x*–*y* table with controller, fiber optics with IR-cutoff and neutral light filters, a spot lens, a linear polarizer, and a rotating analyzer. The choice of optics and CCD camera was made with a view to enabling the entire sample (130 μm thick) to be within the depth of field of the imaging device with the best possible resolution. The *x*–*y* translation table has a travel of 160 mm in both *x* and *y* directions with a resolution (step size) of 0.2 μm, a repeatability of 1 μm, and an accuracy of 6 μm. Transmitted illumination from a halogen DDL lamp is provided through a fiber optic with an attached IR-cutoff filter, a neutral light filter, and a linear polarizer. The light from the fiber optic is focused on the sample using the spot lens. The camera is connected to a computer through the FireWire port, and the *x*–*y*-table controller is connected to the computer through the RS-232 port. A computer program that controls the plate scanning process was written in C^{++}. It allows for plate alignment, for setting up camera parameters, for automatic scanning of all 96 wells, and for storing raw images into an external 500 GB FireWire disk. Imaging and data storage for a 96-well plate takes approximately 4.5 min. Raw images are subsequently batch-processed to perform Bayer interpolation, white balance, and image enhancement. Processed images, stored as JPEG files, identified by plate and well number and scan date, are inspected and scored by skilled personnel using a computer screen (22 inch, 1,600 × 1,200 pixels).

8.3. Crystallization Trials

8.3.1. *In-Meso* Crystallization

In-meso crystallization begins with an initial mixing of the protein solution with the lipid. In this step, the mesophase forms spontaneously and in the case of membrane proteins, the protein is assumed to reconstitute into the lipid bilayer. Homogenization is carried out in a syringe mixer[9] using 60% (w/w) monoolein and 40% (w/w) protein solution, as described.[12] The protein-laden mesophase is then transferred to a 100-μL gas-tight syringe fitted with a 10-mm-long, removable needle. The syringe is mounted on Arm 2 of the robot and its plunger is engaged with the motorized pump mechanism. A 96-deep-well block of precipitant solutions and the 96-well glass plate/spacer combination are put into position on the deck of the robot. The *x*, *y*, and *z* coordinates of the center of the first well in the plate is found by adjusting the position of Arm 2 until the tip of the syringe needle just touches the base plate. This last step takes about 5 min to complete. The same *z*-setting is reliable for a set of at least 12 plates. The coordinates are entered into the robot controller and robotized dispensing of mesophase and precipitants commences.

This happens in the following order. The first four (or eight) wells are loaded sequentially with 50 nL of cubic phase. Arm 2 moves to the park position away from the plate and stabs the syringe needle into a moist sponge. This serves double duty—to prevent dehydration of the cubic phase in the tip of the needle during storage and to clean the tip of any carry-over mesophase. While parked, Arm 1 with its four (or eight) tips is activated to aspirate simultaneously four (or eight) different precipitant solutions from the precipitant block and to dispense 1 μL of each over the cubic phase boluses sitting in the first four (or eight) wells. Arm 1 moves to the wash station where the residual precipitants in the tips are expelled and the inside and outside of the tips are washed and rinsed with Milli-Q water. Arm 1 then moves to its own park position. What has just been described represents a full cycle that is repeated 24 (or 12) times until all 96 wells are filled. The wells are capped with the glass coverslip and a soft rubber brayer is used to provide tight sealing. The entire process of loading and sealing a single 96-well plate takes about 13 min to complete with the four-tip robot. With the eight-tip instrument, the time required is reduced to 8 min. The sealed plate is placed in a temperature-controlled incubator for crystallization to occur.

8.3.2. Screening

The contents of each well in a plate must be examined individually to evaluate the crystallization trial. Currently in the lab this is done using a combination of manual screening and automatic imaging coupled with manual evaluation. The screening schedule adhered to after trials are set up is as follows: Day 1, manual; Day 3, automatic; Day 7, manual; Day 14, automatic; Day 28, manual. Manual screening requires that the researcher examines the plates on a microscope equipped with 4×, 10×, and 40× objectives; a polarizer; and a rotating analyzer. Automated screening makes use of the imaging station introduced above. Images are recorded in normal light and between crossed polarizers, and are inspected and scored by qualified personnel. Observations are recorded into spreadsheets using a 0 to 9 crystallization rating scale developed specially for *in-meso* work. Voice confirmation of numerical scoring and data entry has been implemented. This facilitates the annotation process since the person screening does not have to look away from the microscope and at the computer screen to verify that the correct score has been entered.

8.4. Robot Performance Characterization

Accuracy and precision of the *in-meso* robot as far as delivery volume is concerned were evaluated by direct imaging. Thus, desired volumes of precipitant or lipidic mesophase were dispensed in replicate into wells of a 96-well glass crystallization plate. The plate was sealed with a glass coverslip in such a way that the precipitant solution or the mesophase was sandwiched between two glass plates. The plate was scanned immediately using the imaging robot with 1× and 3× magnification, respectively, in the case of precipitant solutions and the lipidic mesophase. Images were analyzed using Adobe Photoshop or ImageJ (Wayne Rasband, NIH, Bethesda, Maryland; available free from http://rsb.info.nih.gov/ij/) programs to determine an area occupied by the precipitant drop or the bolus of mesophase. Volumes were calculated as the product of the area and the well-spacer thickness (130 μm). The cubic phase boluses used for quantitative analysis were not covered with precipitant solutions and had a uniform disc shape after it was compressed between the two glass plates. In addition, the average dispensing volume was calibrated by checking the actual movement of the syringe plunger after 100 deliveries of a fixed volume.

8.5. Crystallization Plates

The plates were designed in-house for *in-meso* crystallization of membrane proteins. As noted, they have been adapted for other types of crystallizations.[10] With regard to *in-meso* approach, the protein is delivered to the crystallization well dispersed in a lipidic mesophase. Optically, these can present problems especially at the mesophase/aqueous boundary that invariably develops surface roughness. The rough surface scatters light and interferes with examination of the interior of the mesophase bolus, which is where the crystals usually form. The system is somewhat forgiving in the case of highly colored proteins such as bR. However, with small colorless crystals, surface roughness can mean that potential screen conditions are missed because the crystals go undetected. The problem can be overcome if the mesophase bolus is physically sandwiched between two optically clear surfaces. This eliminates the mesophase/aqueous medium interface and the corresponding roughness. The clear surfaces considered for the current application include glass and plastic plates. Plastic was eliminated because, without exception, all plastics have some degree of birefringence, which compromises optical quality. Thus, glass plates were chosen and were found to perform extremely well (see Fig. 2 in the article[10]). With glass there is no problem due to birefringence and submicroscopic crystallites can be seen when the wells are examined between crossed polarizers (Fig. 2g in the article[10]). This is vital information that identifies conditions for subsequent optimization trials.

The glass plates that are used as a base for the crystallization wells have other features that make them particularly useful for robotic *in-meso* crystallization. These include flatness (less than ±10-μm variation across the plate diagonal) and thickness (±50-μm variation in thickness between plates) uniformity. As will be pointed out below (**Section 8.6.1**), the distance between the needle tip dispensing the cubic phase and the base of the crystallization well is critical to uniform and complete delivery of the cubic phase. Too big a distance and delivery does not occur. Too small a distance and delivery is incomplete. Thus, it is crucial that the correct distance be determined and that it is maintained while the plate is being loaded. Since the dispensing height of the syringe tip can be maintained within fairly tight limits by the robot itself, the flatness of the upper surface of the crystallization plate ultimately determines the effectiveness or otherwise of the delivery process. The plate sits on an aluminum block, which, in turn, resides on the deck of the robot. The block has been machined and shimmed to produce a flat surface. Because the glass plates are uniformly flat and thick, when they are placed on top of the block, successful and reproducible delivery in all 96 wells is essentially guaranteed. Thus, glass serves very well in this capacity. In addition, the adhesion between glass and the lipidic mesophase is quite strong that facilitates delivery.

The crystallization plates were designed also with a view to making the imaging process as efficient as possible. With manual evaluation, typically the full volume of a sample is examined by continuously adjusting the depth of focus on the microscope. This can be implemented with the automatic imager as well. However, it takes time. In the interests of eliminating the need to record images at different depths, the field depth of the imager optics was chosen to match sample thickness. The latter was dictated by the thickness of the polyester spacer available commercially. Thus, with the instrument properly focused, a single image is all that must be recorded to obtain an in-focus record of everything in the sample.

While the plates have many desirable properties, including the fact that the glass base and coverslip are reusable, they do suffer one major drawback. They are limited in that they can really only be used for screening purposes where conditions are identified for crystal growth. The next step in the process is to harvest the crystal and to mount it in preparation for diffraction data collection. This is not particularly easy in these plates given that the upper and lower windows are made of glass both of which become glued to the spacer. Since one of the plates is very thin cover glass and is extremely fragile, the likelihood of successfully separating the components cleanly is very low. It might be that a glass-cutter could be use for the purpose. One other possibility is to replace the upper coverslip with plastic. This could then be removed by cutting, with a scalpel for example, and the contents of the well exposed for crystal harvesting. The plastic would bring with it undesirable birefringence as already noted and would have to be suitably thick to provide hermetic well sealing. The approach taken in the lab currently is to use the plates for screening purposes only and to use 72-well microbatch plates to grow crystals for harvesting. Leads found with the robot have been reproduced manually. Typically when a successful condition is identified in an initial screen, an optimization grid screening is performed with the robot. Optimized conditions are scaled up subsequently and performed manually in plates suitable for crystal harvesting. At this stage, a small adjustment in the concentration of the principal precipitant has been found to be useful.

After loading a plate, the final step involves capping with a glass coverslip. A brayer is used to produce a tightly sealed compartment within each well. However, since samples are stored for several weeks to months, water loss from the plate is a concern. While not quantified systematically for the current study, we have found that the plates sealed in the usual way and stored at 20°C do not succumb to noticeable dehydration over a 12-month period.

The base plate employed in the original version of the crystallization plate was used untreated and as supplied by the manufacturer. However, it was found that with some low surface tension precipitant solutions, particularly those containing alcohols, there was a tendency for the drop to spread on the plate and away from the bolus of cubic phase. To counteract this, the plates were silanized. This corrected the problem. All of the base plates used for crystallization trials are now silanized. The adherence of the protein/lipid mesophase to the glass surface does not appear to be compromised in any way by this treatment.

8.6. Crystallization Robot

8.6.1. Dispensing the Lipidic Cubic Phase

The robot described herein is unique. What sets it apart from other robots is its focus on membrane protein crystallization by the *in-meso* method. This requires that the robot is capable of delivering accurately and reproducibly nanoliter volumes of highly viscous lipid/protein dispersion. What makes this possible is the inclusion in the robotic device of a positive displacement microsyringe with a motor-driven plunger and precise motorized positioning of the dispensing needle tip with respect to the well base. However, the robot would not be able to function as such had we not been able to provision the dispensing syringe with relatively large volumes of homogenous lipid/protein dispersion. This was made possible by implementation of a mechanical mixing device developed in the lab for a related application.[9] The robot described here achieves all of the above specifications and is now in routine use in the author's laboratories at the University of Limerick (Ireland) and at The Ohio State University.

As might be expected, the performance of the robot is critically dependent on the reproducibility and accuracy with which it delivers the lipid/protein dispersion. Since the volumes and masses involved are really small and the speeds at which things happen are high, it is not simple to quantify delivery with a view to identifying factors that impact on the quality of the delivery. To assist us in identifying such factors, we recorded the process using a video camera with the assistance of a 12× zoom lens. Stills from the video recordings show the steps in the delivery process (**Plate 8.5**). An examination of such data indicated that the following factors influence the delivery process: dispersion composition and volume, needle-tip-to-well-base distance, needle-tip profile and diameter, and the time profile of the delivery cycle. In our experience, the most important of these was the height of the dispensing needle tip above the well base. When the height is too great the bolus of cubic phase emerges from the tip in the form of a tube that, as it grows, curls back on itself and away from the base of the well. In such an instance, there is no delivery. When the distance is too small, the bolus of cubic phase balls up around and sticks to the tip. Then, when the syringe is withdrawn it takes with it part or all of the dispensed bolus. Again, delivery fails. We have examined systematically the effect of tip-to-well distance on delivery and have established that, for the bulk of the materials we have worked with so far, a distance of between approximately 150 and 300 μm works well for dispensing volumes of 50 nL. We have not drawn up tables of the dependence of optimum distance on the assorted variables because of the complex relationships that exit. These factors are coupled to varying degrees and what we do in practice is to perform a quick test to establish optimum distance for each new system we encounter. In addition to highlighting the importance of the dispensing height distance, this also serves to emphasize the need for accurate and precise *z*-positioning of the robot's Arm 2 and for a flat receiving surface as provided by the glass base plate.

An examination of the video footage of the dispensing process revealed an interesting property of the viscous lipid/protein dispersion that impacts on effective delivery. This has to do with the fact that upon activating the syringe plunger to expel a set volume of the dispersion, the mesophase emerges from the needle tip at a rate that is slower than that expected from the plunger movement. Thus, there is a delay for complete elimination from the needle tip of the volume specified by moving the plunger. Depending on the system and the volumes involved, this can extend to several seconds. An indication of this is apparent when frames **c** and **d** in **Plate 8.5** are compared. We have not investigated the origin of the effect. It presumably has to do with the compressibility and relaxation rate of the dispersion, which may include a small amount of trapped air. Initially, we considered the delay to be a problem. The sense was that we would need to wait at each well for complete expulsion of material before advancing to the next well. The time overhead of this would have made the delivery process unacceptably long (approximately 20–25 min to complete a 96-well plate). However, we realized that provided the time for complete expulsion is less than the well-to-well duty cycle of the robot, then a steady-state would be achieved. In this case, each well will receive the allotted dispersion volume with the exception of the first few wells where presteady-state conditions prevail. Thus, in the interval when the needle tip is being moved out of one well (Well A) and into position in another (Well B), relaxation is happening and material is exuding from the needle tip. Relaxation is complete by the time the tip is in place in Well B. Accordingly Well B receives "carry-over" from Well A plus the amount dispelled in the time the tip is in place in Well B. In essence, the volume that is lost to carry-over by a given well is made up by carry-over from the preceding well. To avoid presteady-state underfilling of wells for the first few dispenses, the syringe is "primed" by expelling a small amount of the dispersion prior to loading the first well on the plate. We have evaluated the dispensing process for a standard volume of 50 nL cubic phase and the results are presented in **Fig. 8.1**. The data show that delivery has an associated standard deviation of within 10% of target volume and that steady-state is already established in the second well.

The smallest volume that can be dispensed using the current apparatus is 20 nL. However, the corresponding reproducibility is considerably less than with 50 nL with a standard deviation of 20% (**Fig. 8.1**). This is one of the reasons we have chosen to work with 50 nL as the standard volume with which to perform the bulk of our crystallization trials. Other reasons for working with the larger volume include the reduced effect of dehydration and the higher probability of crystallization.[10]

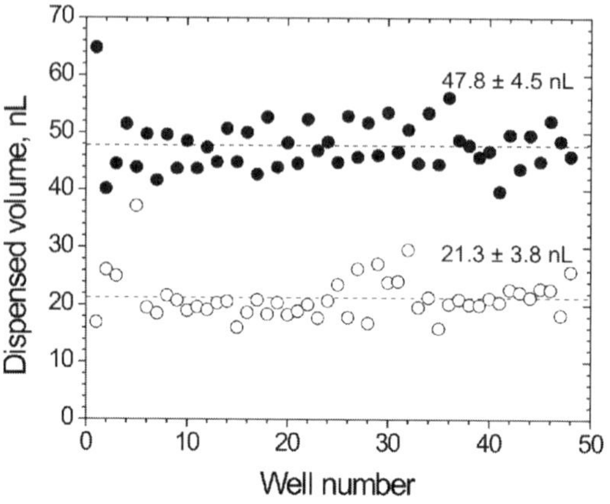

Figure 8.1. Precision and accuracy of the mesophase dispensing component of the *in-meso* robot. Repetitive measurements were performed using 20-nL (open circles) and 50-nL dispensing volumes (solid circles). Actual volumes dispensed were determined as described under "Robot Performance Characterization" (**Section 8.6**). The average and standard deviation for each dispensed volume are indicated.

8.6.2. Dispensing the Precipitant Solution

As noted, the robot has a liquid handling arm (Arm 1) and a lipidic mesophase handling arm (Arm 2). To ensure accurate dispensing of small volumes of precipitant solutions each of the four or eight liquid handling tips on Arm 1 was calibrated gravimetrically. Calibration was done with 1-μL volumes since this is the precipitant volume in use for the bulk of the crystallization trials currently. Thus, 1 μL of water was dispensed ten times into a small, light-weight (500 mg) plastic tube and weighed using a microbalance. Weight was converted to volume using a density of 0.99823 g/mL (20°C, *CRC Handbook of Chemistry and Physics*). The procedure was repeated five times. Based on the average volumes delivered, a calibration factor was input into the program controlling the liquid handler to move each tip in the direction of the targeted volume. The procedure was repeated and for all four (or eight) tips a dispense volume within 5% of nominal was achieved.

With the dispensing volume calibration completed, it was necessary to adjust the liquid-handling parameters for the different types of precipitant solutions to be used. This was done with water, 25% (v/v) MPD and with a 25% (w/w) PEG-4000 solution to cover the range of viscosities likely to be encountered in a typical screening. The liquid-handling parameters included aspiration and dispense speed and delays, air gap volume, waste volume, and transport air volume. The values arrived at for the current configuration include aspiration and dispense speeds of 50 μL/s, 0 s delays, 5-μL air gap volume, 1-μL waste volume and 0-μL transport air volume. With these optimized parameters a coefficient of variance (CV) of less than 10% was obtained typically (tip 3 has a higher CV; **Table 8.1**, see footnote to the table).

Table 8.1. Accuracy in dispensing 1 μL drops of water and 25% (v/v) MPD and 25% (w/w) PEG-4000 solutions on a glass slide under optimizing conditions. The data represent statistics on at least 24 dispenses with each tip.*

	Water, μL	CV, %	MPD, μL	CV, %	PEG 4,000, μL	CV, %
Tip 1	1.18	8.2	1.14	8.8	1.12	9.7
Tip 2	0.94	6.9	0.9	6.8	0.95	8.6
Tip 3	0.84	14	0.73	7.9	0.86	13.3
Tip 4	0.95	8.2	0.92	8.4	0.99	5.7
Tip 5	1.05	8.7	1.03	7.4	1.12	6.4
Tip 6	1.16	7.7	1.19	5.2	1.19	6.8
Tip 7	0.89	8.9	0.98	6.8	0.86	7.4
Tip 8	1.0	7.6	0.98	7.2	0.99	9.3

* The data shown here represent a work in progress. Obviously, calibration coefficients for Tips 1, 3, and 6 would require additional adjustment before the tips could be used with confidence in crystallization trials and/or screen solution preparation.

8.6.3. Dehydration during Setup

Under standard operation conditions with the four-tip instrument, it takes about 13 minutes to fill (with 50 nL cubic phase and 1 μL precipitant solution) and to seal a 96-well plate. This means that the first well is exposed to the atmosphere for a period in which evaporation/condensation can occur and for the contents of the precipitant solution to change. The length of exposure and attendant changes lessen with well number on the plate. Naturally, the severity of the problem depends on the composition and volume of the precipitant solution and the relative humidity and temperature of the ambient environment. Our standard conditions include a 1 μL precipitant solution volume and room temperature (21 to 23°C). Relative humidity in the room where the screens are set up is not tightly regulated and it varies from a low of around 10% in the winter to a high of approximately 70–80% in the summer. Since humidity is so variable, we quantified its effect on the changes undergone by 1 μL of water in place of the precipitant during the course of a typical plate loading process. The results (**Fig. 8.2**) show that evaporative loss is significant, amounting to 30% at 55% RH. We consider this level of loss to be unacceptable and set about reducing it by raising the RH of the environment in the immediate vicinity of the plate. This was done using an ultrasonic humidifier, the output of which was directed in two streams in the general direction of the plate. With this new arrangement, providing equal or greater than 85% RH, the maximum loss was reduced to less than 10% that was considered acceptable. Local humidification of the type described is now a part of our standard crystallization screen setup.

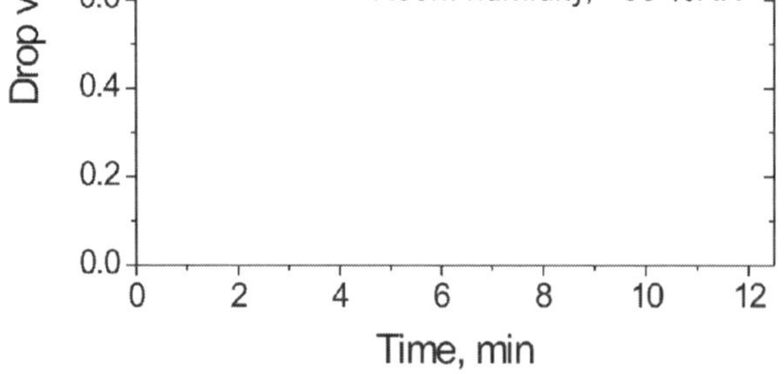

Figure 8.2. Dehydration of 1-μL water drops during the course of a typical 96-well filling operation by the *in-meso* robot. Throughout the filling process, the wells were uncovered and exposed to the atmosphere. Filling was carried out at 20°C with (greater than 85% RH; open circles) and without local humidification (approximately 55% RH; solid circles). Data points at long and short times correspond to wells on the 96-well plate that were filled first and last, respectively. Drop volume was determined as described under "Robot Performance Characterization" (**Section 8.6**).

8.7. Imaging Station

8.7.1. Performance Characteristics

The primary purpose of the image station is to record a high-quality image of the contents of the crystallization well. The emphasis is on image quality that determines what will and will not be seen in the visual inspection of the image. It ultimately defines the success of the overall screening process. We had already in place a standard set of conditions for crystallization and a plate in which to perform the trials. The plates have unequalled optical properties since the critical elements are made of glass. The specifications on the imaging system therefore were to provide the highest-quality image without the need for sampling along the optical axis of the sample and to image a large enough area of the well to include the bolus of cubic phase. The latter should accommodate a bolus 700 μm in diameter and for positional variability of the bolus in the imaging (x,y) plane as dictated by Arm 2 of the robot.

One of the critical elements of any imaging device is the lens. Both resolution, R, and depth of field, DF, depend on the numerical aperture, NA, of the lens, in the following way: $R = 0.61\,\lambda / NA$; $DF = n\,\lambda / NA^2$, where λ is the wavelength of light ($\lambda \approx 500$ nm) and n is the refractive index ($n = 1$ for air). NA is defined as the sine of the angle between the marginal ray and the optical axis of the lens multiplied by n.

After testing several lenses we settled on a 12× zoom lens as a good compromise between lens characteristics, image quality, and price. Continuous zoom provides the ability to adjust magnification, which, of course, is tied to field of view and numerical aperture. Our standard imaging protocol is to scan each plate with a 3× zoom. At this magnification, $NA = 0.051$, $DF \approx 190$ μm, and $R \approx 6.0$ μm. Thus, the entire content of the well, which is 130 μm thick, is in focus with a resolution that is good enough to see micrometer-sized crystals in a bolus that is approximately 700 μm in diameter. The field of view of the 1/2" CCD sensor that is used in recording is 2.4 × 1.8 mm^2 that allows for some variability in bolus positioning in the imaging plane. This represents an image with 2,048 × 1,532 pixels.

The type of illumination used is critical to the performance of the imaging system. We have observed that the diffuse light employed by many commercial manufacturers of crystallization imaging equipment provides very low contrast between small colorless crystals and the lipidic cubic phase in brightfield images. Focusing the light increases contrast but also brings out unwanted features and defects in the mesophase such as cracks and droplets. In our experience, optimal performance is achieved using Koehler-type illumination with a variable numerical aperture condenser. Here, the divergence of the light can be optimized on an individual sample basis. As a low cost alternative to a condenser, a spot lens installed at the end of the fiber optic (**Plate 8.4**) provides adequate image quality in most situations.

The imaging system is required to image all wells in a crystallization plate. This is done either by moving the imager or by moving the plate. We have chosen the latter option using a high-precision *x–y*-translation stage upon which to mount the plate. The stage has a positional accuracy of 6 μm that is considerably better than the accuracy of Arm 2 of the robot (200 μm). Accordingly, it is the latter that limits the overall performance of the imager. As noted, the area imaged has been increased to compensate for this positional variability in sample positioning.

Speed is of the essence in any high-throughput robotic system. Thus, the rate at which the imaging can be executed is critical to overall throughput. This has two principal components: (1) the rate at which the translation stage moves from well to well and (2) the time to transfer image into storage. With the current arrangement, a 96-well plate can be scanned and images stored in 4.5 min.

8.7.2. Image Processing

Digital images of individual wells on each crystallization plate are recorded using a CCD camera in a 16-bit unprocessed format. In the interests of time, each raw 2,048 × 1,532 pixel image is stored on a computer directly and processing is done later. It consists of Bayer interpolation, white balance adjustment, and image enhancement. The latter is implemented using Adobe Photoshop Actions in batch mode and involves corrections of "white" and "black" points, shadow recovery, increasing contrast, and image down-sampling (to 1,024 × 768 pixels) and sharpening. Special Photoshop Actions have been optimized to process the following image types: colored protein images, colorless protein images, and images recorded between crossed polarizers.

With these image-processing features in place, crystals larger than 10 μm are usually easy recognized. A selection of processed images is shown in **Plate 8.6**.

Formation of droplets and other inhomogeneities in the lipidic mesophase are not infrequent events. They can significantly obstruct the view and can give rise to false positives where crystals are identified incorrectly (**Plate 8.6b**). However, the ambiguity can be resolved in the case of birefringent crystals by examining the corresponding image recorded between crossed polarizers (compare panels **b** and **c** in **Plate 8.6**).

8.8. Crystallization Trials

By way of demonstrating the performance of the new *in-meso* robot, crystallization trials have been performed with membranc proteins, bacteriorhodopsin (bR), light-harvesting complex 2 (LH2) and cobalamin (vitamin B_{12}) transporter BtuB, and with the water soluble, lysozyme. The results are included in **Plate 8.7**. Trials included the following commercially available screens: Crystal Screen HT, Index HT, Wizard I and II, SaltRx, and JBScreen, representing a total of 480 conditions, and were performed in duplicate for each protein.

Fig. 8.3 shows the progression in the number of hits observed in crystallization trials of bR and lysozyme. It is interesting to note that one-day post setup approximately one third of the total number of hits is already identified and that no new hits appear after approximately 18 d and 28 d in case of bR and lysozyme, respectively.

8.9. *Quo Vadis ?*

The instrumentation just described is now in routine use in our laboratory. However, there are elements of both the robot and the imaging sta-

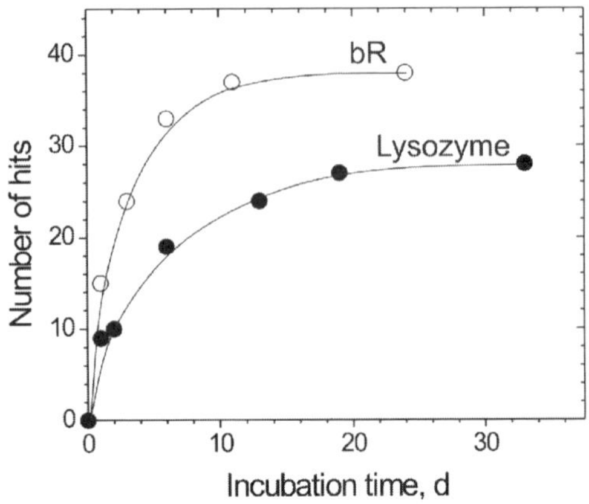

Figure 8.3. Progress in crystallization of bacteriorhodopsin (open circles) and of lysozyme (solid circles) in the monoolein cubic phase using the *in-meso* robot. For both proteins, a total of 480 conditions were screened in duplicate. A hit was registered when an identifiable crystal (usually equal or greater than 5 μm in maximum dimension) was observed in the well. In the case of bacteriorhodopsin, submicrometer-sized "crystals" that appeared as dot-like birefringence were also included in the list of hits. These were particularly prevalent under conditions close to those that yield readily identifiable crystals.

tion that could benefit from improving. In the first instance, it is desirable to speed up the crystallization-plate loading process so that samples have less time to succumb to evaporative loss. As noted, the current filling time is about 13 min with the four-tip robot and 8 min with the eight-tip instrument. We expect to be able to reduce these times significantly (20–25%) with more efficient robot-controlling programs and by effecting simultaneous arm motion.

It is imperative that finding the *x*, *y*, and *z* coordinates of Well A1 of a given plate is done with great care and accuracy because the accuracy with which the rest of the plate is filled depends on it. Currently, this is done manually and takes some time. We would like to automate the process and we have proposed the use of what is called a "soft-sensor" for coordinates measurement.[13] Briefly, this involves using a small bolus of cubic phase at the dispensing needle tip to register when the tip is close to contacting the base of the well. This is a work in progress and has not been settled on as the route to follow and other options are being examined.

As noted, the relative humidity of the environment in which screens are set up has a profound effect on the final composition of the screen.

Currently, we control very roughly the local environment of the plate in a way that does reduce evaporative loss to an acceptable level. We can and will do better by enclosing the robot in a windowed chamber where RH will be regulated in the 90 to 95% range under computer control.

The current plate/robot configuration produces a disc-shaped bolus of mesophase. The rate at which equilibration with the crystallant solution ingredients occurs and the exact time profile of these events during the course of the crystallization trial depends on location within the bolus. A greater range of conditions within a given well would be realized were it possible to tailor-make bolus shape. Given the current setup, this is limited to discs, elongated slabs and wedges. While we have not experimented with shape as a variable with regard to crystallization behavior, we have been able to produce boluses having an assortment of shapes. This is one area worthy of investigation that might improve overall crystallization success rate.

Crystals, once they form, must be harvested and mounted for diffraction measurements. These are slow steps that are currently done manually. We are exploring ways of automating both processes and some preliminary results have been reported.[14] To effect automatic harvesting with the *in-meso* system however will require that we also overcome the difficulty of crystal recovery from within the glass-windowed wells currently in use.

Once filled and capped, plates are placed in temperature-controlled incubators. They are then evaluated manually and at the image station on a defined schedule. Moving toward higher throughput will necessitate reliance on the image station completely and on automatic transfer of plates between incubator and imaging station. An industrial grade robot is being considered for the latter application.

The manual scoring of crystallization trials is a definite bottleneck in the current setup. We are working on automatic image recognition and scoring programs optimized to evaluate images of *in-meso* crystallization trials taken with the imaging station. This task is much more challenging than evaluation of crystallization trials of soluble proteins setup using conventional methods since the lipidic mesophase often develops imperfections and/or droplets that have sharp edges and facets. Images taken using cross-polarizers will be used to aid in crystal identification in those cases.

Bar coding for plate identification and tracking has not yet been implemented but will be shortly. Bar-coded labels will be attached to the plates (**Plate 8.3a**) and will be identified and logged by means of a reader fitted to the image station. It will be imperative to have the bar-coding system in place when the robotic plate manipulator to move plates between the incubator and the image station is installed.

Finally, to handle and to mine efficiently the wealth of data associated with a crystallization project such as this, a comprehensive database framework is needed. Accordingly, we are in the process of designing and implementing a relational database for archiving crystallization trial-related data. This includes everything from an inventory of lipids, proteins, and crystallants used in trials all the way to scores of the trials themselves. We would also like to implement a web-based data-mining interface to the above database to support more rationally designed crystallization trials based on an analysis that includes conditions producing hits as well as misses.

8.10. Conclusions

A high-throughput robotic system has been developed for crystallizing membrane proteins using lipidic mesophases. It incorporates commercially available components and is relatively inexpensive. The crystallization robot uses standard automated liquid handlers and a specially built device for delivering accurately and reproducibly nanoliter volumes of highly viscous protein/lipid mesophases. Under standard conditions the robot uses just 20 nL protein solution, 30 nL lipid, and 1 μL precipitant solution. 96 wells can be set up using the four-tip robot in 13 min and in 8 min with the eight-tip model. Trials are performed in specially designed 96-well glass plates. The slim (less than 2 mm high) plates have exquisite optical properties and are well suited for detection of microcrystals and for birefringence-free imaging between crossed polarizers. Quantitative evaluation of crystallization progress is done using an automated imaging system. The optics, in combination with the slim crystallization plates, enables in-focus imaging of the entire well volume in a single shot such that a 96-well plate can be imaged in just 4.5 min. The crystallization robot is also used for screen solution preparation.

References

1. Caffrey M. **Membrane protein crystallization.** *J Struct Biol* 2003, **142**:108–132.
Review on *in-meso* crystallization technique.

2. Nollert P, Qiu H, Caffrey M, Rosenbusch JP, and Landau EM. **Molecular mechanism for the crystallization of bacteriorhodopsin in lipidic cubic phases.** *FEBS Lett* 2001, **504**:179–186.
Mechanism of *in-meso* crystallization of bacteriorhodopsin.

3. Faham S and Bowie JU. **Bicelle crystallization: A new method for crystallizing membrane proteins yields a monomeric bacteriorhodopsin structure.** *J Mol Biol* 2002, **316**:1–6.
Bicelle crystallization.

4. Cherezov V, Yamashita E, Liu W, Zhalnina M, Cramer WA, and Caffrey M. ***In meso* structure of the cobalamin transporter, BtuB, at 1.95 Å resolution.** *J Mol Biol* 2006, **364**:716–734.
Crystals of the outer membrane β-barrel protein BtuB grown *in-meso*.

5. Schobert B, Cupp-Vickery J, Hornak V, Smith SO, and Lanyi JK. **Crystallographic structure of the K intermediate of bacteriorhodopsin: Conservation of free energy after photoisomerization of the retinal.** *J Mol Biol* 2002, **321**:715–726.
Highest-resolution structure obtained by the *in-meso* method.

6. Paas Y, Cartaud J, Recouvreur M, Grailhe R, Dufresne V, Pebay-Peyroula E, Landau EM, and Changeux J-P. **Electron microscopic evidence for nucleation and growth of 3D acetylcholine receptor microcrystals in structured lipid-detergent matrices.** *Proc Nat Acad Sci USA* 2003, **100**:11309–11314.
Microcrystals of acetylcholine receptor—bungarotoxin complex grown *in-meso*.

7. Chiu ML, Nollert P, Loewen MC, Belrhali H, Pebay-Peyroula E, Rosenbusch JP, and Landau EM. **Crystallization *in-cubo*: General applicability to membrane proteins.** *Acta Cryst* 2000, **D56**:781–784.
Five different membrane proteins are crystallized *in meso*.

8. Raman P, Cherezov V, and Caffrey M. **The membrane protein data bank.** *Cellular and Molecular Life Sciences* 2006, **63**:36–51.
Web-based data bank of membrane protein structures.

9. Cheng AH, Hummel B, Qiu H, and Caffrey M. **A simple mechanical mixer for small viscous lipid-containing samples.** *Chem Phys Lipids* 1998, **95**:11–21.
Syringe-based device for mixing lipid and protein solution for *in-meso* crystallization trials.

10. Cherezov V and Caffrey M. **Nano-volume plates with excellent optical properties for fast, inexpensive crystallization screening of membrane proteins**. *J Appl Cryst* 2003, **36**:1372–1377.

Glass-based plates for effective screening of *in-meso* crystallization trials.

11. Cherezov V, Peddi A, Muthusubramaniam L, Zheng YF, and Caffrey M. **A robotic system for crystallizing membrane and soluble proteins in lipidic mesophases**. *Acta Cryst* 2004, **D60**:1795–1807.

Automation of *in-meso* crystallization.

12. Misquitta Y and Caffrey M. **Detergents destabilize the cubic phase of monoolein: Implications for membrane protein crystallization**. *Biophys J* 2003, **85**:3084–3096.

Procedure for manual *in-meso* crystallization setup.

13. Muthusubramaniam L, Peddi A, Zheng YF, Cherezov V, and Caffrey M. **Automating crystallization of membrane proteins by robot with soft coordinate measuring**. *Proc IEEE International Conference on Robotics and Automation*, New Orleans, Louisiana, April 2004, 2:1450–1455.

Soft-sensor for measuring coordinates in *in-meso* robot.

14. Chen W, Peddi A, Zheng YF, and Caffrey M. **Automating crystal harvesting and mounting for high-throughput macromolecular crystallography**. *Proc World Congress on Intelligent Control and Automation,* Hang Zhou, China, June 2004, **5**:4650–4655.

Automation of crystal harvesting.

9

Automated Classification of Crystallization Experiments

Julie Wilson

9

Automated Classification of Crystallization Experiments

Julie Wilson

University of York, York, United Kingdom

9.1. The Need for Automated Classification

The use of robots has led to a dramatic increase in the number of crystallization experiments performed in most laboratories; and, in structural genomics centers, tens of thousands of experiments are produced on a daily basis. Visual inspection of the results is becoming increasingly impractical and the development of tools to automate analysis is essential. Imaging systems are now available to record the results of crystallization trials at regular intervals and software to automatically analyze the images is now being developed by a number of research groups.

The fundamental aim is the detection of crystals but reliable classification of other outcomes would provide information for subsequent trials. In the absence of crystals, the occurrence of microcrystals, spherulites, or sea urchins indicate conditions that are very close to those required for crystal growth and conditions resulting in phenomena such as phase separation or crystalline precipitate may also be refined in optimization protocols. Experiments resulting in heavy amorphous precipitate or denatured protein on the other hand show that the conditions are inappropriate. Experiments may therefore be ordered so that only the highest scoring images need be checked by eye.

This chapter presents the major step in the analysis and classification of images from crystallization experiments and describes some of the approaches taken by research groups producing software for automated analysis. Edge detection methods, fundamental to many image analysis techniques, are introduced in **Section 9.2**. Just as we are able to recognize an object from a rough line sketch, so the computer program uses the edges in an image as the most important features in pattern recognition. Identification of the crystallization drop boundary then allows the analysis to be restricted to the appropriate area of the image. This speeds up processing and is vital to cope with high-throughput robotic systems. First, the boundary of the well is identified and the image cropped accordingly, as described in **Section 9.3**. Before attempting to define the drop boundary, it may be worth eliminating unwanted artifacts that could affect the mask obtained for the drop. The removal of objects that could cause problems in classification is discussed in **Section 9.4** and the masking of the crystallization drop in **Section 9.5**. At this point, characteristic features that can be quantified and used for classification must be determined. These may be features of the whole crystallization drop or of individual objects within the drop and should ideally have similar values within a class and different values between classes. Feature extraction is discussed in **Section 9.6**. Finally the different classes are considered in **Section 9.7** and the most commonly used classification methods are described.

9.2. Edge Detection

The edges of objects give rise to sudden changes in intensity from one pixel to the next and can therefore be identified by analyzing the gradient, or rate of change, of the image intensity. At each pixel, we need to know the maximum magnitude of the gradient and the direction in which that gradient occurs. It is simplest first to estimate gradients in two perpendicular directions, usually the x and y axes of the image, and then combine them to give the total gradient. Mathematically the gradient is calculated by differentiation but simple operators can be used to approximate the gradient of an image and they form the basis of many edge-detection methods. The Sobel operator used by Wilson[1] approximates the rate of change in x (the horizontal direction), using the filter shown in **Fig. 9.1b**.

a

x_1	x_2	x_3
x_8	x_0	x_4
x_7	x_6	x_5

b

-1	-2	-1
0	0	0
1	2	1

c

-1	0	1
-2	0	2
-1	0	1

Figure 9.1. The eight immediate neighbors, $\{x_1, \ldots, x_8\}$ of the pixel x_0 are shown in (**a**). (**b**) and (**c**) show the Sobel filters for x and y respectively.

That is

$$G_x \approx (x_7 - x_1) + 2(x_6 - x_2) + (x_5 - x_3)$$

for each pixel, x_0, where $x_1, \ldots, x_8$ are the pixel's immediate neighbors as shown in **Fig. 9.1a**. Similarly, the rate of change in y (the vertical direction) is approximated by

$$G_y \approx (x_3 - x_1) + 2(x_4 - x_8) + (x_5 - x_7),$$

using the filter shown in **Fig. 9.1c**. The magnitude of the gradient, M, giving the steepest change at x_0, is calculated from the equation

$$M = \left[G_x^2 + G_y^2\right]^{1/2},$$

and the direction of this change is determined by the angle, α,

$$\alpha = \tan^{-1}\left(\frac{G_y}{G_x}\right)$$

measured with respect to the x-axis. The gradient magnitudes shown in **Figs. 9.2** and **9.3** and the gradient angles in **Fig. 9.4** have been calculated with the Sobel operator.

To locate the important details, some threshold must be specified for a significant gradient magnitude. Taking all pixels with magnitudes

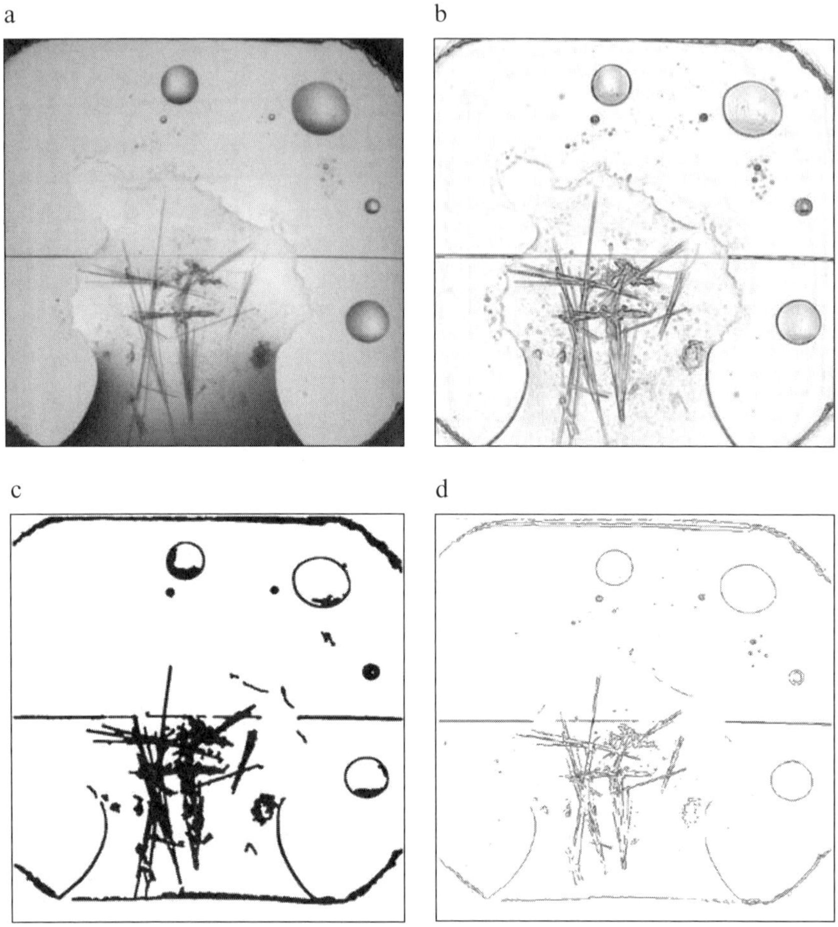

Figure 9.2. The image in (**a**) has a great deal of background noise and the edge of the drop is not very clear. The gradient magnitudes for this image are shown in (**b**).

above this threshold allows objects to be defined as shown in **Fig. 9.2c** and only considering the local maxima of these pixels defines the edges in the image, as shown in **Fig. 9.2d**. The Sobel operator identifies objects quickly and simply but is not ideal for detecting edges. Other more complicated algorithms allow weak edges to be followed where they are connected to strong edges as well as avoiding those due to noise. Canny's algorithm for edge detection[2] is by far the most popular method and has been used in the analysis of crystallization images.[3,4]

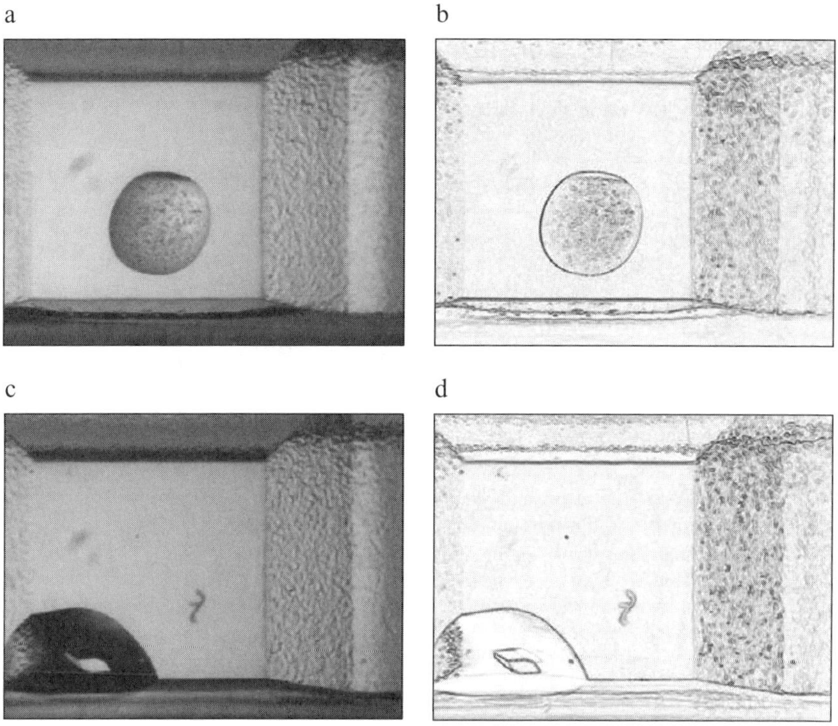

Figure 9.3. (**a**) and (**c**) show crystallization drops in Greiner low-profile trays with the corresponding gradient magnitudes in (**b**) and (**d**) respectively. The drop can overlap the sloping sides of these wells as seen in (**c**), creating problems for classification. In the gradient magnitudes the effects of the plastic, seen through the drop in (**d**), look very similar to the genuinely interesting contents of the drop in (**b**).

Other methods use the rate of change of the gradient, or Laplacian operator, to detect edges. More details can be found, for example, in Gonzalez and Woods.[5] Cumbaa et al.[6] use an approximation to the Laplacian to detect image discontinuities in crystallization images and squared Laplacian values to measure the local smoothness of an image.

To avoid problems caused by shadows at the edge of the drop, Bern et al.[4] select different directions for their initial gradient estimates and introduce the directional gradient. For each pixel they compute the gradients away from the drop center and perpendicular to this. They then set the outwards gradient to zero if its sign indicates normal shading, i.e., light to dark, before estimating the total gradient.

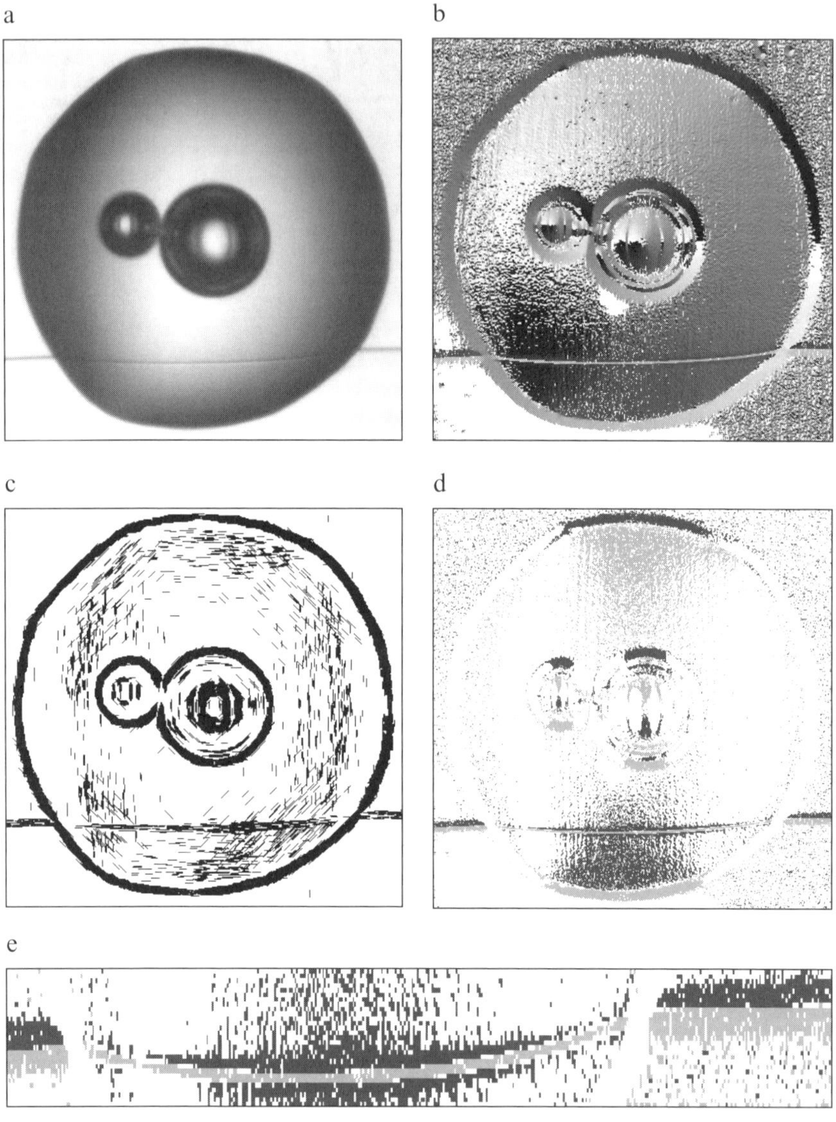

Figure 9.4. The crystallization drop shown in (**a**) is empty in terms of the experiment but the bubbles and the horizontal line are likely to create problems for automated classification. Both are clearly visible in the gradient angles shown in (**b**). In (**c**), those pixels that lie in a straight line with three pixels on either side all having very similar gradient angle are shown in black. The gradual change in angle allows circular objects to be identified easily. In (**d**) all pixels for which the gradient angle is close to 90° are shown in light gray and those close to 270° are shown in dark gray. (**e**) shows a close-up near the horizontal line showing the curvature of the line and that the pattern has missing pixels.

9.3. Image Cropping

As **Fig. 9.3** shows, the crystallization drop may be relatively small in comparison to the size of the image. Robotic dispensing and movement during storage and imaging make the position of the drop impossible to anticipate. A considerable area of the crystallization tray must be captured in order to ensure that the well, and therefore the drop, is entirely contained within the image. The edges of wells with straight vertical sides and expected separation for a particular experimental setup are easily recognized in the gradient images. This is not so straightforward in wells with sloping sides such as those shown in **Fig. 9.3**. Whilst the horizontal edges are obvious, it can be seen that the effects of the plastic mold and the lighting have blurred the vertical edges, particularly on the left of the image. Although squiggles in the plastic do not give sharp straight edges, there is a noticeable difference in the pattern between the sloping sides and the flat bottom of the well. The vertical sides can be identified by low variation in intensities along a row for an expected number of pixels based on the size of the well and the resolution of the image. This means that the crystallization drop does not affect the identification of the well boundaries even where it overlaps. Summing over these rows for each column of pixels allows the vertical edges to be located. Comparison of the magnitudes inside the drop in **Fig. 9.3b** with that part of the drop overlapping the well sides in **Fig. 9.3d** shows that the pattern caused by the plastic would make accurate analysis of this part of the drop very difficult and it is probably better to restrict the analysis to that part of the drop that does not overlap the well sides.

Cropping the images down to the size of the well removes up to two-thirds of the original pixels and therefore significantly reduces the time required for image processing. This is particularly important in high-throughput centers where tens of thousands of images are produced daily.

9.4. Artifacts not Related to the Crystallization Experiment

The image in **Fig. 9.2** has already been cropped, but much of the information in the gradient is still related to the area outside the crystallization drop. The aim is now to find a mask for the drop and so restrict the analysis to the relevant part of the image. The small droplets and the horizontal line, an artifact due to the way the plastic trays are made, seen in **Fig. 9.2** would both create problems when identifying the boundary of the crystallization drop. Furthermore the line in the plastic tray seen within the drop is extremely likely to create false positives in classification, as straight lines are also obvious characteristics of crystals. Such objects provide no information about the experiment and should if possible be deleted before the mask for the drop is generated.

As the gradient angle gives the direction of steepest gradient, its direction is always perpendicular to edges in an image. It will be constant along a straight-line edge and will change gradually around smooth curves. In most cases this allows the line in the plastic tray and small circular droplets to be identified and eliminated prior to analysis.

9.4.1. Bubbles and Droplets

Along with additional droplets, such as those seen in **Fig. 9.2**, high-speed dispensing by robots can also create air bubbles in the crystallization drop (see **Fig. 9.4a**). Again these are likely to be overrated in classification as reflected light gives the bubbles a high variation in intensity usually associated with more interesting objects. In **Fig 9.4b** both the edge of the crystallization drop and the bubbles are clearly visible in the plot of the gradient direction from the gradual change in angle. Circular droplets and bubbles can both be identified from the obvious concentric circles of smoothly changing gradient angles. The pixels that have sufficient number of neighbors with similar gradient angle can be identified as shown in **Fig 9.4c**. Centers of circular objects can then located by counting the number of these pixels situated on circles of varying radii. This allows larger droplets and bubbles to be identified as well as the edge of the crystallization drop if it is circular. There could be confusion between rounded crystals and smaller bubbles and therefore only objects with a radius between preset minimum and maximum values should be deleted at this stage.

9.4.2. Plastic Mold Effects

Fig. 9.4b clearly shows the constant gradient angle along the horizontal line. These lines are created where molten plastic meets during the manufacture of the trays and are usually roughly horizontal or vertical. This helps provide a method for their identification. Along a horizontal line we expect to find angles of 90° closely followed by angles of 270° and similarly, along a vertical line we expect to find angles of 0° closely followed by angles of 180°. In **Fig. 9.4d**, only those angles very close to 90° (light gray) or 270° (dark gray) are shown for the image in **Fig. 9.4a**. Although the line seems obvious, the close-up in **Fig. 9.4e** shows that there may be pixels "missing," an effect caused by the line being out of focus. Furthermore the crystallization drop has a lens effect that distorts the line as it passes through the drop. This distortion is dependant on the curvature of the drop and can be quite severe in viscous drops. Allowances can be made for these effects but must be balanced with the risk of eliminating the edges of crystals that happen to lie horizontally or vertically across the drop. Crystals however can be expected to lie within the drop, whereas these lines usually extend beyond the crystallization drop and this fact can be exploited. Small, disconnected sections should not be masked in order to reduce the risk of deleting crystals. Instead these should rely on the classification algorithms for identification.

9.5. Masking the Drop

With the image cropped to the size of the well and any artifacts not relevant to the experimental outcome removed, the area to be analyzed should ideally be restricted to the crystallization drop. This involves identifying its limits, an easy task if the drop is globular as shown in **Fig. 9.4**. In this case the strong lighting effect due to the curvature of the drop gives a sharp change in the grayscale and the drop boundary would therefore show clearly in the gradient magnitudes. **Fig. 9.4b** shows that the boundary is also obvious from the gradually changing angle in the plot of the gradient direction. This is not the case for the image in **Fig. 9.2** where the drop has spread across the well and the gradient magnitude has low values along the boundary, making a suitable mask for the drop more difficult to obtain.

9.5.1. Circular Masks

In most cases the crystallization drops are globular and a reasonably fitted circular mask is easily generated. **Fig. 9.5a**, for example, shows a mask generated by the method described in Wilson.[1] The image can now be reduced again from the size of the well to the diameter of the circle. However, such a mask is obviously unsuitable for drops that have been disturbed and can create problems even when the drops appear to be more or less circular. **Fig. 9.5** also shows drops for which the best-fitting circle would mask areas of the drop potentially containing objects of interest. As well as the danger of losing information, a badly fitting mask

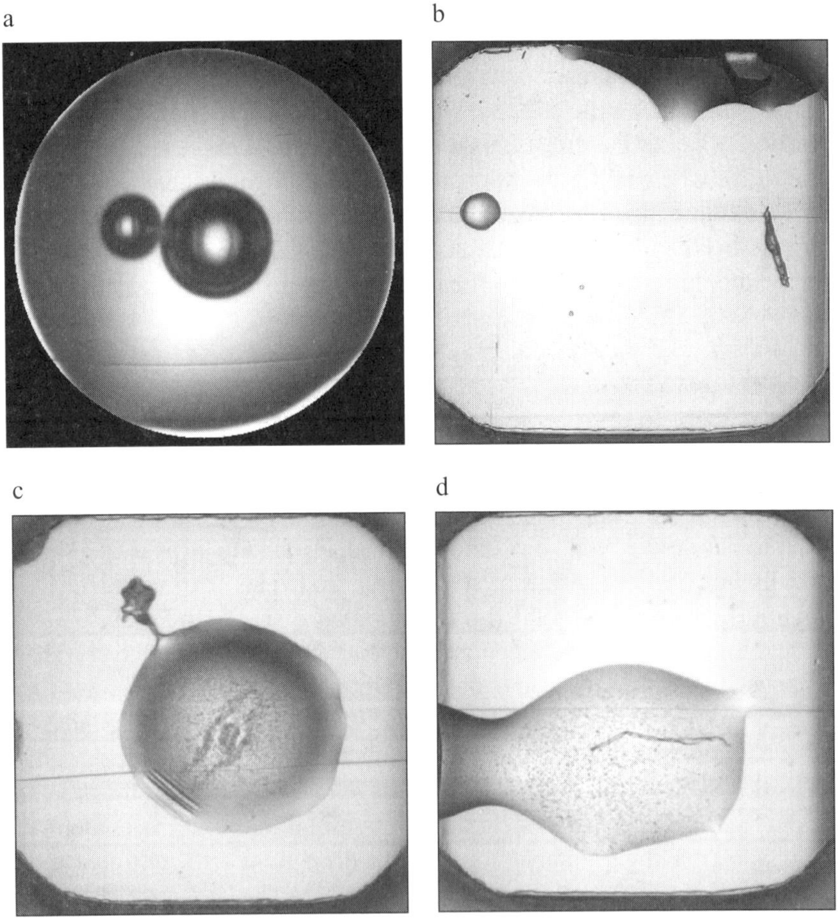

Figure 9.5. (**a**) shows that the circular mask obtained for the image in **Fig. 9.4a** fits well. However a circular mask would be of no use when the drops are disturbed as seen in (**b**). A circular mask could be fitted to either of the drops shown in (**c**) or (**d**) but would exclude some potentially interesting areas.

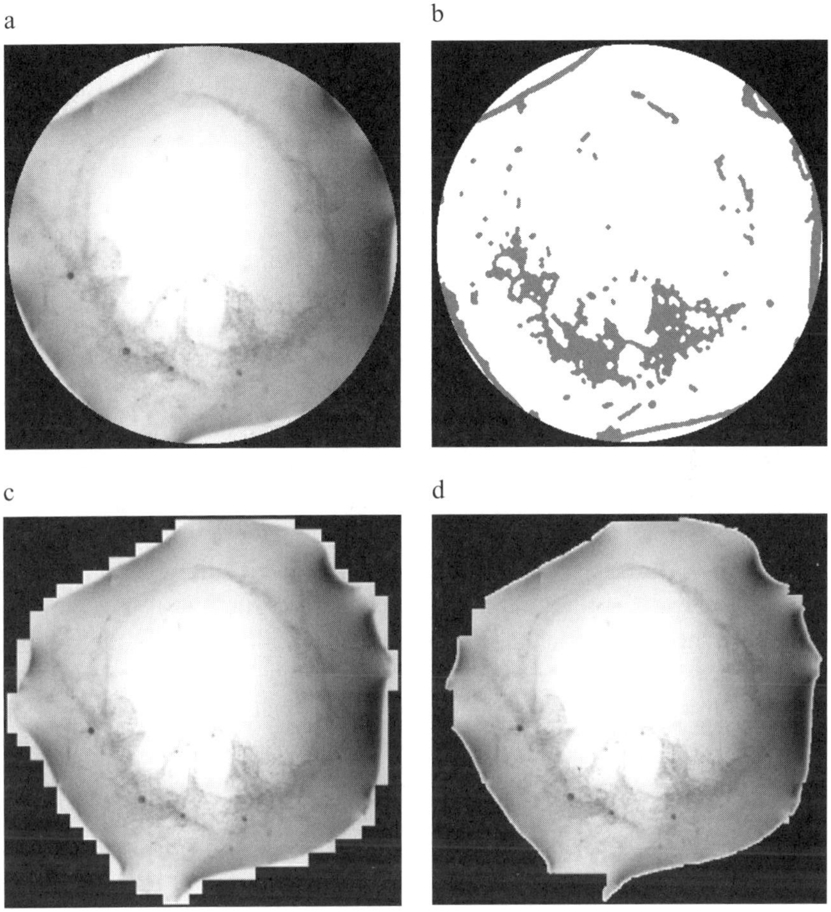

Figure 9.6. A crystallization drop for which a circular mask fits well is shown in (**a**) and the objects obtained for this image are shown in gray in (**b**). The visible sections of the drop boundary create objects that may lead to false positives. (**c**) shows the mask obtained by considering the gradient angles on a coarse grid and (**d**) shows this mask after refinement.

may also create objects likely to give rise to false positives in classification. In **Fig. 9.6** the circular mask eliminates most of the background without losing much of the crystallization drop but the remaining drop boundary can still cause problems particularly if the drop is to be analyzed in terms of distinct objects.[1,7] The straight edges, as well as sharp changes in intensity due to light and shadow, are characteristics used to identify crystals. As a continuous drop boundary is more easily identified than separate sections, it helps to make the radius slightly larger than necessary for the mask. When considering individual objects, short boundary segments can often be identified by their proximity to the edge of the mask in combination with other attributes.

9.5.2. Non-Circular Masks

Cumbaa et al.[6] use a probabilistic model to segment the well into three regions on a coarse grid. Each coarse pixel is assigned to be empty well (W), inside the drop (D) or on the edge of the drop (E). They then exclude the coarse pixels identified as either W or E from the analysis. This avoids the problem with the drop boundary creating false positives but risks the loss of crystals growing at the edge of the drop.

Another method to provide an initial mask for the crystallization drop on a coarse grid uses the gradient direction to determine the status of each coarse pixel.[7] As there is little change in the grayscale along some of the drop boundary in **Fig. 9.2**, the magnitude of the gradient here is small. However, as the direction of the steepest gradient is still perpendicular to the edge of the drop, it is more easily identified by the smooth change in gradient angle. The variation in angle between neighboring pixels can be used to identify coarse pixels that the edge of the drop contains. **Fig. 9.6c** shows a mask obtained in this way. Rather than eliminate the "edge pixels" it is possible to refine the mask within them once the object pixels have been obtained. **Fig. 9.6d** shows the mask after refinement.

9.6. Analysis of the Crystallization Drop

Automated classification of images requires measurable features of the various classes to be determined and exploited. For example, straight lines are an obvious characteristic of crystals and the number of these and their significance can be quantified, potentially in a single variable. Of course noncrystalline objects may also have straight edges whereas not every crystal does and several classification variables describing different features need to be evaluated for discrimination between classes. Ideally each attribute would have similar values within classes and different values between classes. For each object to be classified, whether it is an object within the crystallization drop or the drop itself, a feature vector consisting of the classification variables is evaluated and used to assign the object to a particular class. Some of the features used in the analysis of crystallization drops are discussed below.

9.6.1. Feature Extraction

Any pattern-recognition problem requires features to be extracted from the input data that can be used in some pattern-matching or decision-making procedure for classification. These features should be measurable characteristics that are common within classes and discriminatory between classes. In order to classify the images from crystallization experiments, characteristics that can determine different classes must be extracted from the images. These may be properties of the whole drop or of individual objects within the drop.

9.6.2. Object-Based Features

Whilst sudden changes in the grayscale give the steepest gradient, the gradient has some value at every pixel. Taking those pixels for which the gradient magnitude is above a prespecified threshold provides a method of identifying individual objects within the drop. After some tidying to remove odd pixels, each connected set of pixels can be considered as an object to be evaluated separately in terms of a number of different attributes. Each property needs to be quantified in terms of a single variable so that objects can be assigned to different classes based on these values. The objects to be considered have no predefined shape or size and there is a great deal of overlap in the values of the variables expected for the different classes but each feature gives some information about the object in question.

9.6.2.1. Straight Lines

Straight lines are obvious characteristics of crystals that can be exploited for classification. Straight edges should have similar gradient magnitudes along them as well as fairly constant gradient angles. Rather than only consider the boundary of an object, straight lines in both the gradient magnitudes and directions anywhere within an object can be utilized. With a set number of directions defined from any point, a single variable can be obtained in both cases as follows: From each boundary point of an object calculate the number of pixels in a straight line within the object for which the difference between this value and the previous one is within a predefined limit. Square this number of pixels (to give more weight to longer lines) and keep the "best line" of the possible directions. The average over all boundary points gives a final variable.

9.6.2.2. Shape Descriptors

In **Fig. 9.7**, the object pixels obtained for different types of object show that shape is an important factor in classification. Unlike most single crystals, objects resulting from precipitate generally have long boundaries relative to their overall size, as do clusters of needle crystals. The square of the number of pixels on the boundary divided by the total number of pixels in an object provides a variable that gives shape information. Another variable related to shape is obtained by dividing the number of pixels in an object by the number of pixels in the minimal rectangular box required to completely enclose the object. Compact objects will have higher values than straggly objects such as those arising from denatured protein or skin on the drop. In fact, the shape of the minimal box itself can be used as another classification variable. The quantity is calculated as

$$max(x, y) / (x + y),$$

where x and y are the lengths of the sides of the rectangle, will give a value from 0.5, for a square, to a value close to 1, for a very long thin rectangle.

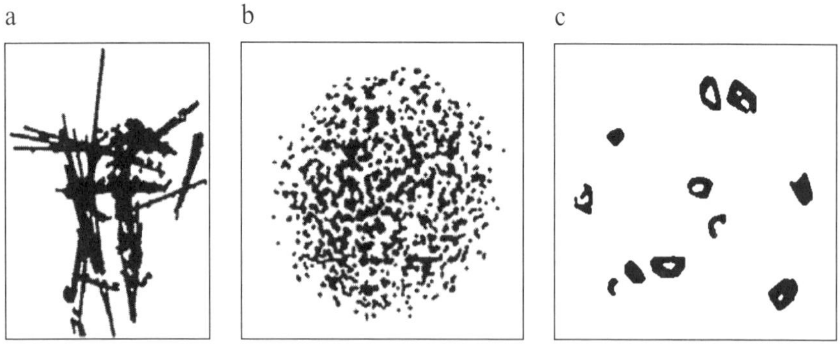

Figure 9.7. The object pixels obtained from the contents of three different drops. Here the three types of object—needle crystals in (**a**), precipitate in (**b**), and single crystals in (**c**)—are recognizable from shape and boundary properties.

9.6.2.3. Statistical Descriptors

Objects arising from microcrystals can look very similar to precipitate in terms of shape but classification variables based on the greater variation in intensity in the image and in the gradient magnitudes can help to distinguish between them. The first and third quartile positions can be determined from the image pixels within each object and used as variables. These should help identify objects with many dark pixels (often a trait of denatured protein) and those with many light pixels (objects that reflect the light).

9.6.2.4. Boundary Descriptors

The way the boundary changes can provide useful information for discrimination between objects. Chain codes provide a representation from which a number of features are easily extracted but require the boundary pixels to be ordered. **Fig. 9.8** shows how the eight possible directions corresponding to a pixel's neighbors can be specified so that the change along a boundary from one pixel to the next can be encoded. Along a straight line, in any direction, the chain code will be constant. For example, the consecutive 7s, 4s and 2s in **Fig. 9.8c** correspond to the straight sections of the boundary shown in **Fig. 9.8b**. Although in practice some prior smoothing is necessary, straight sections of the boundary are easily identified from the chain codes, as the difference between adjacent chain codes (working modulo 8, e.g., $|0-7| = 1 \pmod 8$) will be zero along a straight section. Other differences will be positive or negative according to the convexity or concavity of the curves and have absolute values of either 1 or 2, depending on how sharp the turn is. Thus a measure of curvature can be obtained from these absolute values.

With the boundary pixels ordered, a further shape descriptor can be extracted. The distances between some fixed pixel (the centroid of the object or the first boundary pixel) and every other pixel on the boundary (in order) are calculated and this function of distance used to provide a classification variable. In addition, the distances between the boundary pixels and the centroid of the mask for the crystallization drop give a measure of how close an object is to the edge of the drop. This variable is particularly useful for identifying objects that are actually short sections of the drop boundary and therefore of no interest in reality but may cause false positives in classification.

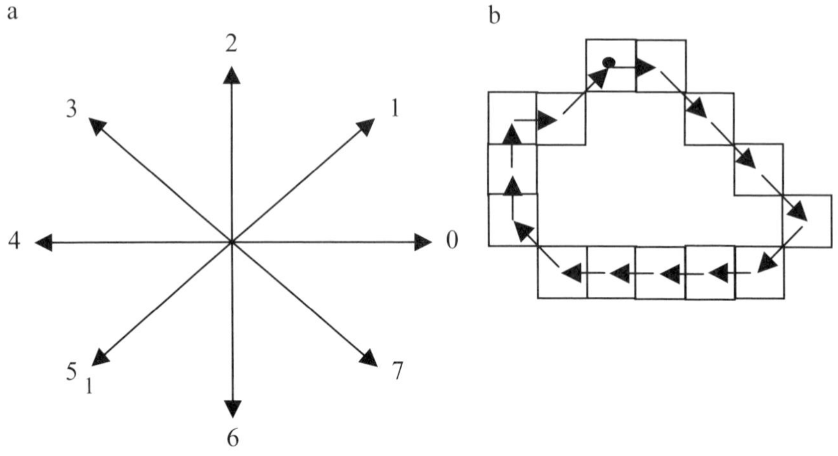

c
0 7 7 7 5 4 4 4 4 3 2 2 0

d
−1 0 0 −2 −1 0 0 0 −1 −1 0 −2 1 −1

Figure 9.8. The eight possible directions showing the change from one pixel to the next along the boundary of an object can be encoded by an integer as shown in (**a**). The chain code is the sequence of integers giving the changes for the entire boundary, in order, from some starting pixel. For example, the chain code for the boundary in (**b**), starting from the pixel with the black dot, is given in (**c**). The differences between adjacent elements in the chain code, shown in (**d**), can provide measures for the straightness and curvature of the boundary.

9.6.3. Drop-Based Features

Various methods exist to evaluate the state of either the entire drop or a block within the drop boundary rather than individual objects within the drop. As well as the presence of straight lines, a number of measures are related to texture defined in terms of the correlation between intensities at different scales. A few global parameters are extracted from many local texture measures and used in classification by Spraggon et al.[3] The ability of Fourier and wavelet transforms to decompose images into different levels of detail mean that they are both commonly used in texture analysis.

9.6.3.1. Straight-Line Transforms

The Hough transform and Radon transform have both been used to identify straight lines within a drop and therefore establish the possibility of crystals. Juristica et al.,[8] determine evidence weights for straight lines from the Radon transform. They then compute a straight-line score by summing weights obtained from all the detected lines considered significant. The maximum weight for the image provides a second variable.

The Hough transform[9] can be used to locate objects of known topology and was used by Zuk and Ward[10] to identify crystal edges. Spraggon et al.[3] and Bern et al.[4] also use the Hough transform to find straight lines in the edges obtained using Canny edge detection. Again a score can be calculated over all straight edges and used for classification along with the maximum straight-edge score for any edge.

9.6.3.2. Fourier Descriptors

Frequency information can be used to derive features for classification. However a two-dimensional Fourier transform can only be computed over a rectangular area and problems arising from the assumption of periodicity need to be overcome. Juristica et al.[8] compute the Fourier transform of the largest square that can be contained within the drop and would therefore miss any crystals outside this area. Bern et al.[4] focus many of their classification features on a small block within the drop boundary. This "best block" is chosen according to the directional gradient so that the risk of missing crystals is minimized.

9.6.3.3. Local Smoothness

Differences in pixel intensities within chosen blocks of pixels provide a measure of the local smoothness of an image. Cumbaa et al.[6] squared the Laplacian values and summed over 16×16-pixel blocks to obtain a smoothness measure for each block. The highest score provides a measure for the image.

9.6.3.4. Statistical Descriptors

Just as with object-based classification, statistical measures calculated within the drop boundary can provide classification variables. For example, Bern et al.[4] use the standard deviation about the mean of the gray levels within the best block.

9.6.3.5. Direct Comparison with Library Images

Scores based on comparison with target images can also be used in classification. This approach is exploited by Cumbaa et al.[6] to identify microcrystals but could be extended to other classes.

9.7. Classification

The range of classes used varies from a simple two-class system, when only the presence or absence of crystals is to be determined, to around ten separate classes. There will always be a class for crystals, or "hits," that may be subdivided to reflect the different possibilities from large single crystals to clusters such as thin plates and needles. Microcrystals often form a further distinct class, as do precipitates, sometimes divided into "good" and "bad" precipitates, and there may be a "junk" class for results such as denatured protein. Finally clear drops must be identified as such and, in order to highlight problems with robotic systems, the event that no drop has been dispensed may be classified separately. However many classes are used, the features extracted should be capable of discrimination through some decision-making process.

9.7.1. Learning Algorithms

In supervised learning algorithms, a network is trained to associate a certain output with particular inputs so that the vector of values obtained from the classification variables, or feature vector, can be used to assign an object (either individual objects within the drop or the drop itself) to a particular class. If separate objects are classified in this way, an overall score for the image needs to be calculated from these results. Unsupervised learning algorithms respond only to the input values and do not require output targets. Such networks are usually clustering algorithms and are useful for the identification of patterns and groupings within the input data. For classification, each node can be associated with a particular class after training, as described for self-organizing maps below.

Whatever the learning algorithm, training should be carried out on a set of input vectors, or training set, that have been preclassified by eye. An independent set of preclassified vectors, not used in training, is required as a test set for validation of the results and to avoid problems associated with over-fitting.

9.7.1.1. Self-Organizing Maps

Kohonen self-organising maps[11] are widely used unsupervised learning algorithms and have been applied to the classification of crystallization images.[3,7] Each node or neuron in a two-dimensional grid has a set of weights, $(w_1, \ldots, w_n)$, associated with it. Here n is the number of classification variables in the feature vector, $(v_1, \ldots, v_n)$. Initially each weight, w_k, is given the average value (over the training set) of the kth variable and the weights are modified over a number of training cycles. In each cycle, every vector in the training set is presented to the network in turn and the node with weight vector closest to the input vector is identified. The weights for this winning node are then adjusted to be even closer to the input vector. In addition the weights of nodes within a neighborhood of the winning node are also adjusted but to a lesser extent. Nodes in this area of the map are now even more likely to win when a comparable feature vector is presented. In this way the map organizes itself so that physically close nodes learn to recognize similar input vectors. After training, the weight vector for each node is compared with each feature vector in the training set and the node is assigned the class of the closest training vector. As each node now has a class associated with it, new vectors can be presented to the map for classification and the class of the node closest to the input vector assigned to it.

9.7.2. Decision Trees

Every node in a decision tree tests the value of some classification variable, or combination of variables, and branches according to the possible values. For example, specifying a threshold provides two branches, one if the variable is above the threshold and one if it is not. Feature vectors are sorted through the tree from a root node, down the appropriate branches to a leaf node that defines the class of the vector.

Bern et al.[4] use a hand-crafted decision tree in which thresholds were set after inspecting scatter plots of the feature values. Trees that take into account the relative cost of various errors can also be generated automatically. Although it has been found that these could be very large and difficult to implement for the classification of individual objects, decision trees can provide a safety procedure when used together with other classifiers.[7] Whereas the distance measure used in neural networks such as self-organizing maps relies on the values of all variables, crucial alternatives in the decision tree can be determined by a single variable. As objects can often be identified by a few particular variables, this difference can be exploited by combining the methods.

9.7.3. Linear Discriminant Analysis

Cumbaa et al.[6] use linear discriminant analysis (LDA) to classify images. Consider two n-dimensional data sets corresponding to two distinct classes, where n is the number of classification features. LDA computes the vector v that maximally separates the points in the two data sets when they are projected onto v. Gaussian distributions are fitted to the two data sets and the distance along v that forms the equiprobability boundary between the two sets is calculated. From this a threshold can be obtained that determines which of the two classes a new feature vector should be assigned to.

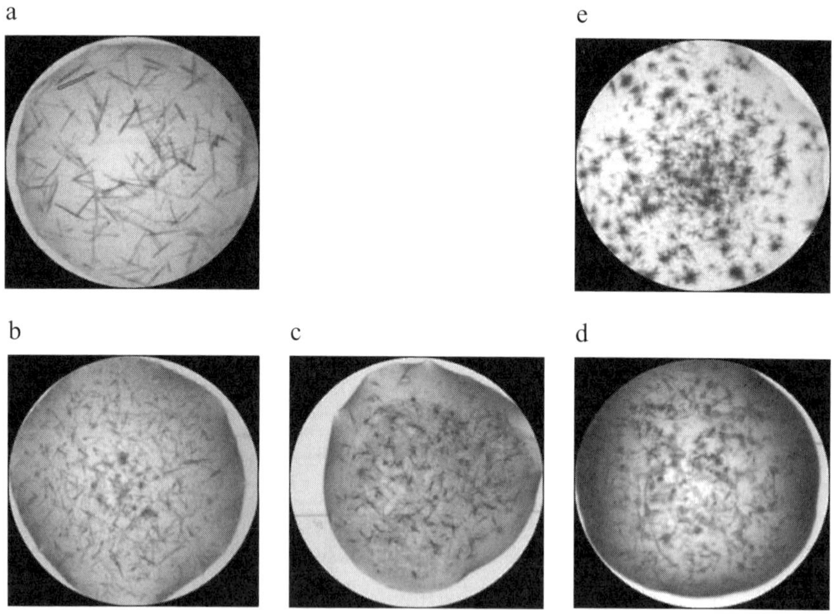

Figure 9.9. The drop in (**a**) contains clusters of needles, some of which may be considered as single crystals. The boundary of one such object is shown. The drops in (**b**) to (**d**) show some intermediate states between this and the fuzzy microcrystals in (**e**).

9.8. Summary

Crystallization experiments do not give results that can be separated easily and form a continuum of states rather than discrete classes. **Fig. 9.9** demonstrates this with some transitional states between microcrystals and needle crystals. Studies on the agreement rates for human classification of crystallization images show that crystallographers agree about the class of an image in around 65% of cases.[12] Considering classification to be correct if there is only one (ordered) class difference increases the agree-

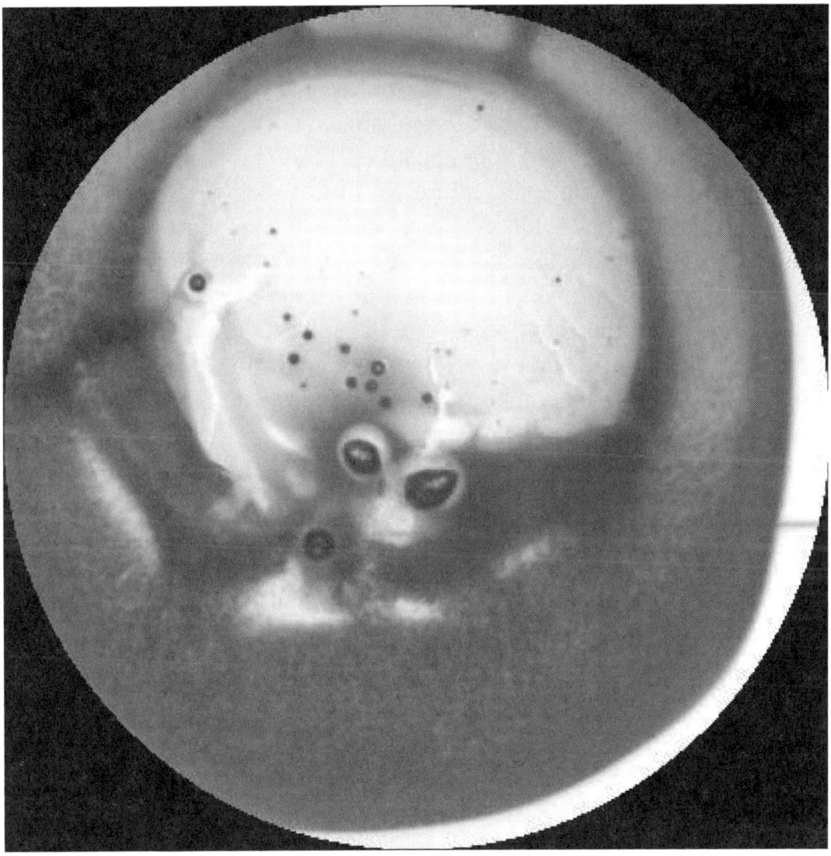

Figure 9.10. The combination of bubbles, heavy amorphous precipitate, and denatured protein give some weird effects. Unfortunately a gremlin in the drop such as this is likely to give a false positive.

ment rate dramatically to approximately 90%. However many discrete classes are used to separate the results of crystallization experiments, there will always be borderline cases and this should be taken into account when assessing the accuracy of automated software. Results should be considered not just in terms of how many "correct" classes are achieved, as two different crystallographers will very often disagree on what is correct, but also by how much an image is misclassified.

For any pattern-recognition problem, the extracted features should have common values within a class and be discriminatory between classes. Unfortunately this is rarely the case with features extracted from crystallization images. Not only is there a great deal of overlap between features from different classes due to the fuzzy class boundaries but the features within a class may have totally unrelated values. Whilst crystals are expected to have straight edges and definite angles, there is no fixed pattern that describes every crystal. Wrinkles in the skin on the drop have similar features and are the cause of many false positives. Denatured protein, another source of false positives, can create an assortment of strange effects such as the "creature" in **Fig. 9.10**.

Most methods for evaluation of crystallization images take the drop-based approach, quite possibly because the task of obtaining a training set for individual objects from all the various classes is extremely tedious. On the other hand, the object-based method of Wilson[1,7] does not take advantage of the overall environment or interrelatedness of the objects. A method to combine the two approaches must surely give improved results.

Whatever improvements are made to the software, human classification will always be superior. However such software promises to drastically reduce the number of images to be inspected by eye. Scoring the images allows them to be examined in order of merit and as soon as some high-scoring conditions are confirmed, no further images need be considered. Equally important is the potential for the development of optimization procedures. Not only will many promising initial conditions be identified but failed experiments will also be recorded allowing the possibility for automated screening protocols to be developed via data-mining and neural-network techniques.

References and further reading

1. Wilson J. **Towards the automated evaluation of crystallisation trials.** *Acta Cryst* 2002, **D58**(part 11):1907–1914.

2. Canny J. **A computational approach for edge detection.** *IEEE Trans Pattern Anal Machine Intell* 1986, **8**(6):679–698.

3. Spraggon G, Lesley S, Kreusch A, and Priestle J. **Computational analysis of crystallization trials.** *Acta Cryst* 2002, **D58**(part 11):1915–1923.

4. Bern M, Goldberg D, Kuhn P, and Stevens R. **Automatic classification of protein crystallization images using a line tracking algorithm.** *J Appl Cryst* 2004, **37**(2):279–287.

5. Gonzalez R and Woods R. **Digital Image Processing.** 2nd ed. Upper Saddle River, NJ: Prentice Hall, 2002.

6. Cumbaa CA, Lauricella A., Fehrman N, Veatch C, Collins R, Luft J, DeTitta G, and Juristica I. **Automatic classification of sub-microlitre protein-crystallization trials in 1536-well plates.** *Acta Cryst* 2003, **D59:**1619–1627.

7. Wilson J. **Automated evaluation of crystallisation experiments.** *Cryst Rev* 2004, **10**(1):73–84.

8. Jurisica I, Rogers P, Glasgow J, Fortier S, Luft J, Wolfley J, Bianca M, Weeks D, and DeTitta G. **Intelligent decision support for protein crystal growth.** *IBM Sys J* 2001, **40:**394–409.

9. Hough P. ***Methods and means for recognizing complex patterns.*** U.S. Patent 3,069,654, 1962.

10. Zuk W and Ward K. **Methods of analysis of protein crystal images.** *J Cryst Growth* 1991, **110:**148–155.

11. Kohonen T. ***Self-Organization and Associative Memory.*** 2nd ed. Berlin: Springer, 1987.

12. Wilson J and Berry I. **Human classification of images from crystallisation experiments.** Private communication.

Index

Data bank(s),

Wave,

Color Plates

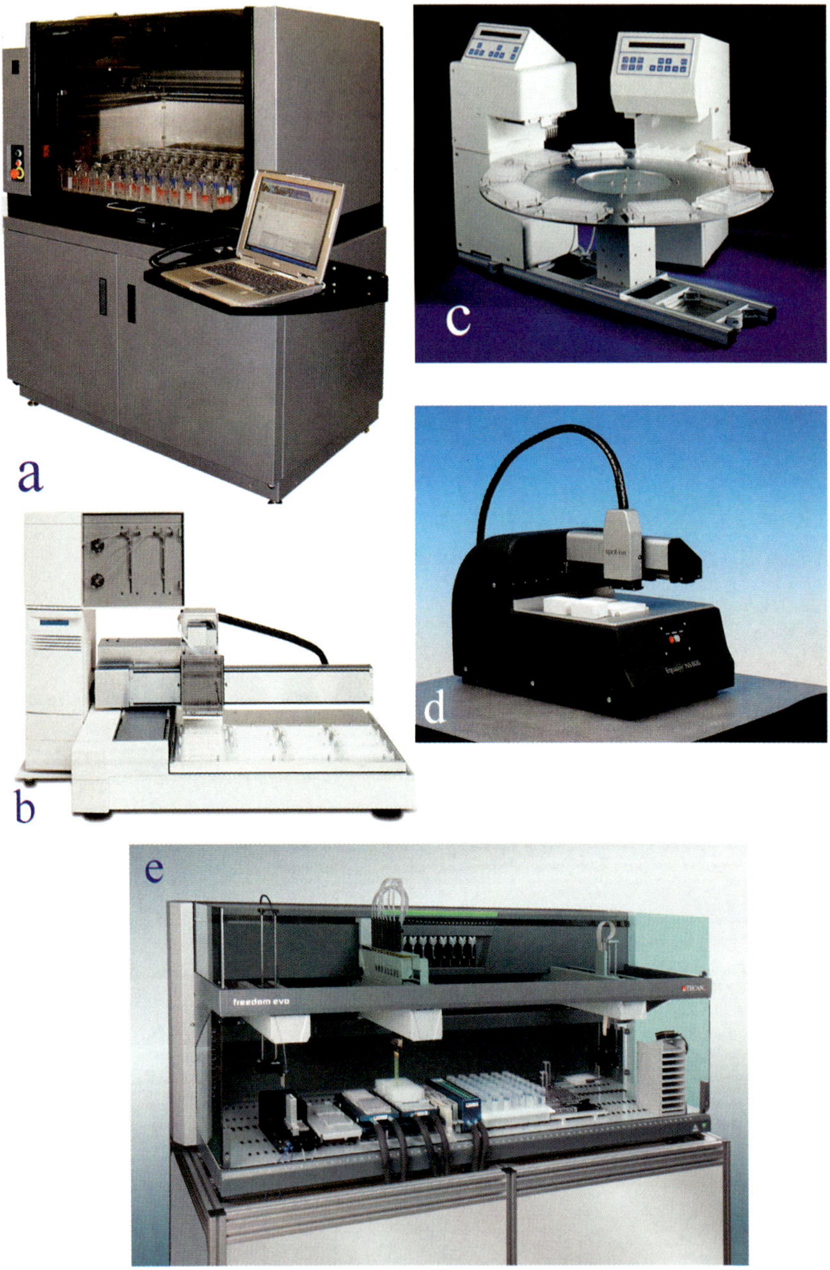

Plate 3.1. (*Continued on next pages*) Commercial crystallization robots.(Arranged alphabetically by the robot's trademarked name.) (**a**) Alchemist™; (**b**) Constellation® 1200; (**c**) CyBi®-HTPC; (**d**) Equator™; (**e**) Freedom EVO®; (**f**) Honeybee 161™; (**g**) Hydra® II Plus-One; (**h**) Matrix Maker™; (**i**) Microlab® STAR; (**j**) Mosquito™; (**k**) Nanodrop™ II; (**l**) Oryx 6™; (**m**) Phoenix™.

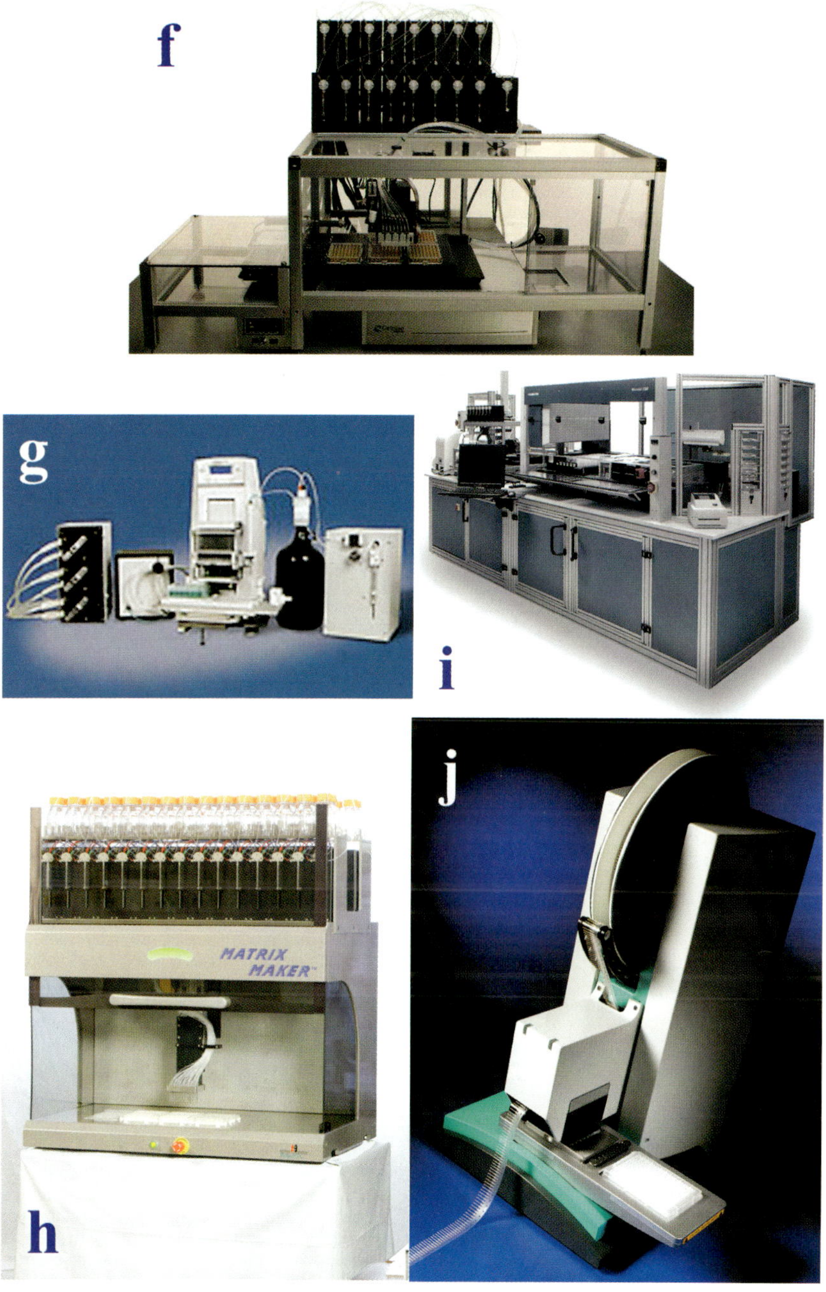

Plate 3.1. (*Continued*)

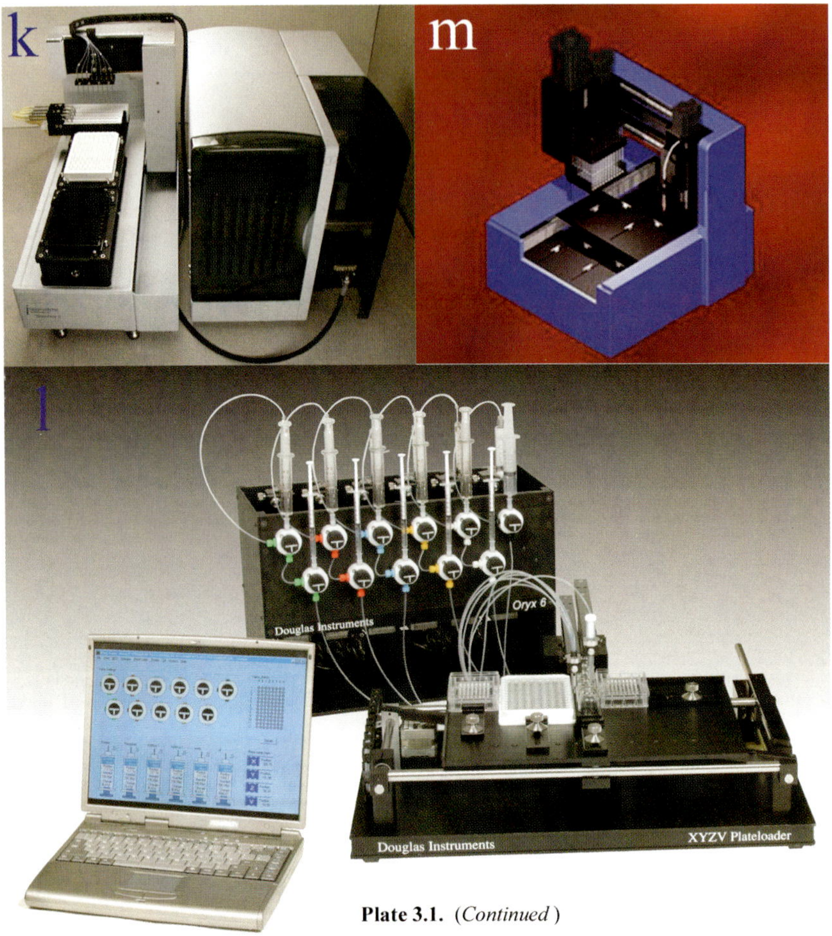

Plate 3.1. (*Continued*)

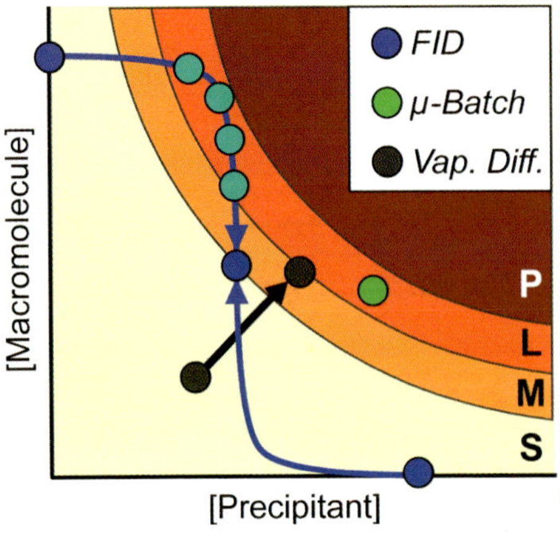

Plate 4.1. Phase-space sampling by traditional crystallization methods. An example microbatch experiment under impermeable oil is shown in *green*, a vapor-diffusion (or microbatch under water-permeable oil) trial in *gray*, and a free interface diffusion run (diagrammed here in the form of a microcapillary setup) in *blue*. The FID experiment traverses a complex, sigmoidal path through space, as solutes equilibrate from regions of high concentration across the length of the crystallization chamber; many areas of the labile/metastable region are sampled over time (*light blue* circles). The vectors for FID and vapor diffusion are representative parametric plots with time as the dependent variable. For FID, the change in concentration over time is diagrammed for only a single point along the length of the FID chamber.

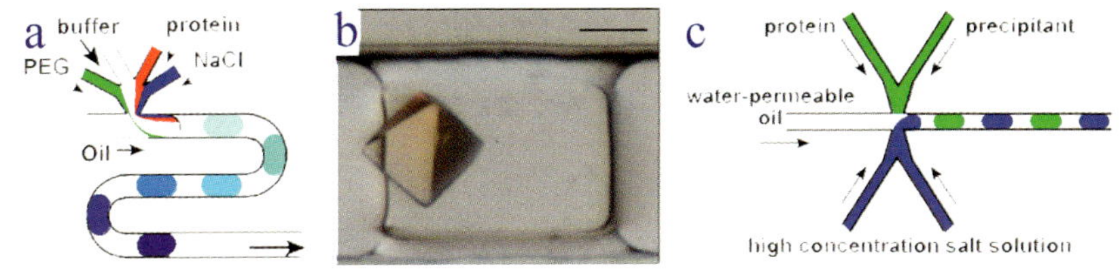

Plate 4.2. Microbatch and vapor-diffusion crystallization in a microfluidics chip. (**a**) Diagram of mixing and equilibration lines in a microbatch chip. Different volumes of four solutions are flowed together into a common chamber, then merged with an immiscible oil to form nanodroplets and allow crystallization to occur. (**b**) Lysozyme crystal grown in a nanodroplet as visualized through the chip. (**c**) Schematic of the mixing and equilibration plumbing lines in a vapor-diffusion chip. Two separate mixing stations are used to create discrete droplets of either sample/precipitant or a high salt solution. The drops are then alternately moved into a channel containing water-permeable oil, but placed adjacent to each other, so as to allow the sample/precipitant mix to dehydrate due to the higher osmotic potential of the high-salt solution. Diagrams reproduced with permission.[59,62]

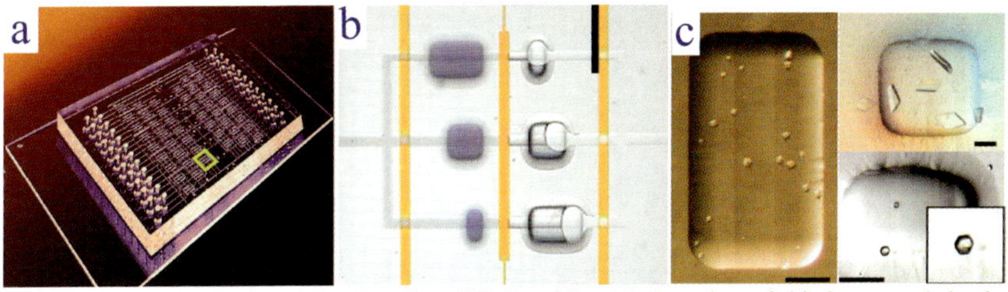

Plate 4.3. Microfluidics-based free-interface diffusion chip. (**a**) Overview of chip layout and circuitry. Reagent input ports (*cylinders*) flank fluid routing lines and crystallization chambers. (**b**) Blowup of a single experimental sector (*green boxed region* in panel **a**) that combines the target solution with a crystallization condition in three different mixing ratios. The chambers are being dead-end loaded (*blue and clear solutions*); containment lines for isolating the chambers and controlling the start of diffusive mixing are filled with yellow dye. Scale bar is 500 μm. (**c**) Nonmodel protein crystals from the author's lab grown in the FID chip. All scale bars are 100 μm—see text for details (pages 97–98). Panels **a** and **b** were reproduced with permission from Fluidigm, Corp. and from the source,[56] respectively.

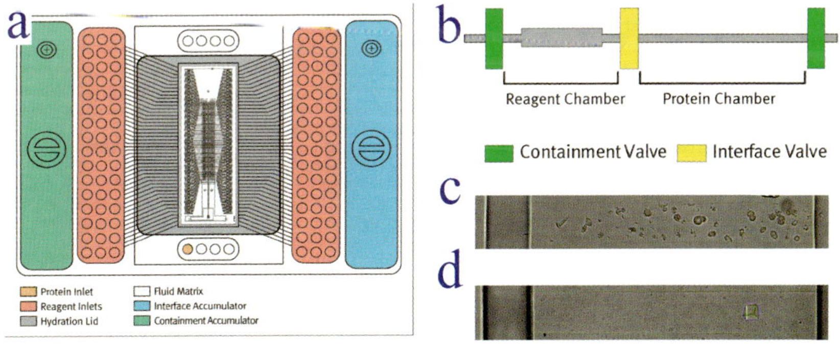

Plate 4.4. TOPAZ™ chip configuration. (**a**) Schematic of a 1.96 chip. The chip occupies the center of the diagram and is connected to 96-reagent wells (*red circles*) by a series of channels (*black lines*). Accumulator chambers (*blue and green rounded rectangles*) eliminate the need for external pressure plumbing to control interface and containment lines. (**b**) Blowup showing the architecture of a single FID chamber in the 1.96 chip. (**c**) Photo of α-lactalbumin crystals growing in the chip. Note the clear distribution of crystal morphology across the length of the chamber as evidence that FID is taking place. (**d**) Image of a single insulin crystal growing in a chip. The width of the chamber (running from the top to the bottom of the panel in this view) is 100 μm. Diagrams reproduced with permission from Fluidigm, Corp.

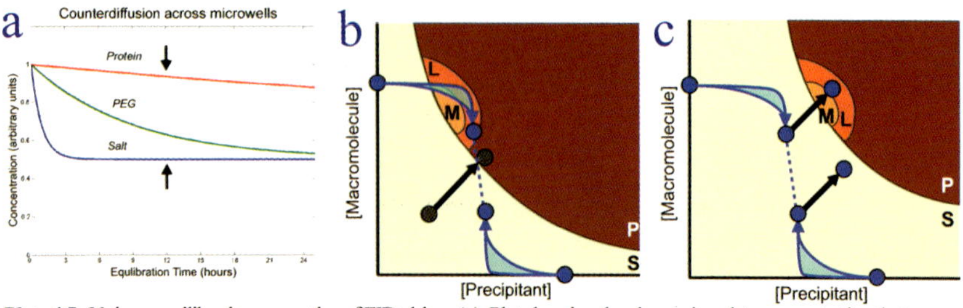

Plate 4.5. Unique equilibration properties of FID chips. (**a**) Plot showing the change in solute concentration that occurs in the nanowells of the Hansen et al. FID chip as a function of time. (**b**) Phase-space coverage by FID chips over a period of time but prior to full equilibration. See pages 100–101 in the text for details. (**c**) FID chips allow further exploration of phase space through dehydration. The diagram is a variation on panel **b**, showing how gentle dehydration of the chips, following closure of the interface valve, can be used to redirect the equilibration paths of the two, now-isolated, sample chambers through an M/L region that occupies a high concentration of both sample and precipitant.

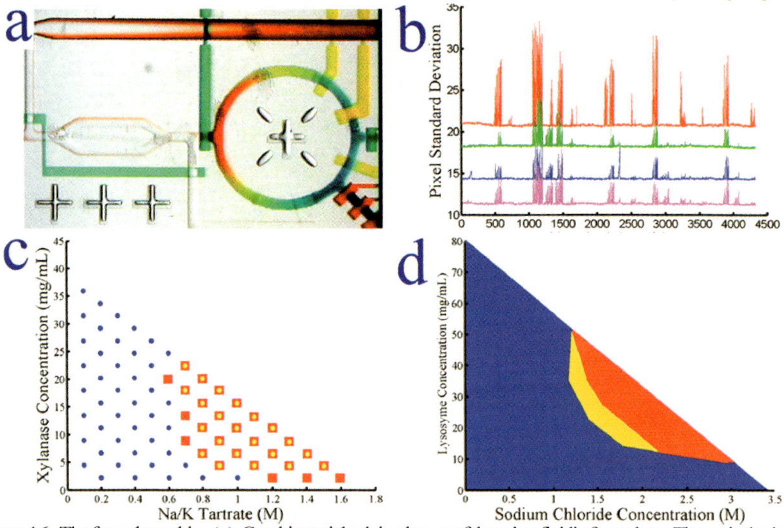

Plate 4.6. The formulator chip. (**a**) Combinatorial mixing layout of the microfluidic formulator. The optical micrograph shows the manipulation of food dyes within the chip. The view is of a sector of the chip that contains the rotary mixer element and its associated actuation valves and sample lines. A color gradient has been formed within in the mixing ring by consecutive injections of blue, green, yellow, and red dyes. These solutions can be mixed together by swirling them around the ring multiple times under the control of a set of three (*yellow*) peristaltic valves. (**b**) Solubility fingerprints of xylanase generated from exposing the protein to dozens of different precipitant, buffer, and salt types over a variety of mixing ratios (~4,300 distinct conditions in total). Each data series represents a separate fingerprinting experiment using the same basis set of chemical conditions. The top solubility fingerprint (*red*) is generated using a sample having elevated protein concentration (90 mg/mL), and exhibits both higher signal to noise and additional peaks not present in the other data series (70 mg/mL). The two center solubility fingerprints were generated sequentially on a single device (first *green*, then *blue*) with the same loaded sample, demonstrating the stability of the protein over the time of the experiment (~40 hours). The bottom solubility fingerprint (*pink*) was generated on a separate device using the same proteinsample as the blue fingerprint, showing reproducibility across devices. Each spike represents a series of conditions conducive to reversible precipitation of the sample. (**c**) Comparison of xylanase phase mapping done both on chip and in microbatch experiments for the precipitant Na/K•Tartrate. Conditions that gave rise to precipitation in microbatch format and in chip are represented by *red squares* and overlaid *yellow circles*, respectively, while those represented by *red squares* precipitated only in microbatch experiments. Conditions that did not produce precipitation in either format are shown as *blue circles*. (**d**) Reversibility of precipitation and solubility hysteresis for lysozyme observed by outward and return titrations from the origin. For the outward titrations, soluble lysozyme was mixed with increasing amounts of precipitant until precipitation occurred. For the return titrations, lysozyme was precipitated at a high concentration of precipitant, then exposed to lower precipitant concentrations until the precipitate dissolved. The *red region* shows the area of phase space in which precipitation was observed in both outward and return titrations. The *yellow region* is the area of hysteresis in which the protein remained soluble for the outward titrations but precipitated in the return titrations. The *blue region* is the area in which the protein was soluble for both the outward and return titrations. Figures reproduced with permission.[67]

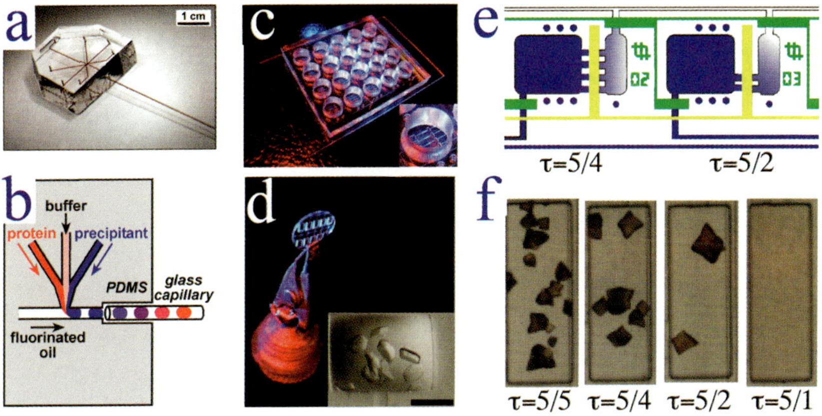

Plate 4.7. Scale-up and diffraction chips. (**a**) Microbatch diffraction chip layout. The PDMS chip is shown attached to a capillary extending from its center. Fluid lines are filled with red dye. (**b**) Schematic of the channel layout of the chip shown in panel **a**. (**c**) Thin layer, FID-based "growth chip." A view of the chip is shown with 20 discrete chambers ("osmotic baths") that can be each differentially dehydrated to control nucleation by layering different amounts of oil on top of the well. Five FID wells, of a type similar to that shown in **Plate 4.3**, are located within each osmotic bath chamber (**inset**). (**d**) Thin layer PDMS disk excised from the master chip and mounted on a microclip attached to a magnetic base for diffraction studies. (**Inset**) Blowup of large, single crystals of the Rho transcription termination factor78 grown from one of the FID nanowells contained in a disk. Scale bar is 200 μm. (**e**) Schematic of the growth chambers from a prototype TOPAZ™ "X-ray" chip. (**f**) Snapshot of catalase crystals growing in sample wells similar to those diagrammed in panel **e**. Figures reproduced with permission from articles[62] (panels **a** and **b**) and[63] (panels **c** and **d**), and Fluidigm, Corp. (panels **e** and **f**).

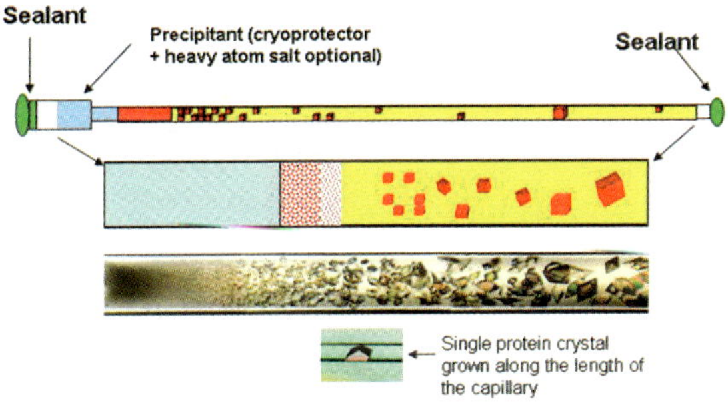

Plate 5.1. Counter-diffusion crystallization under restricted geometry.

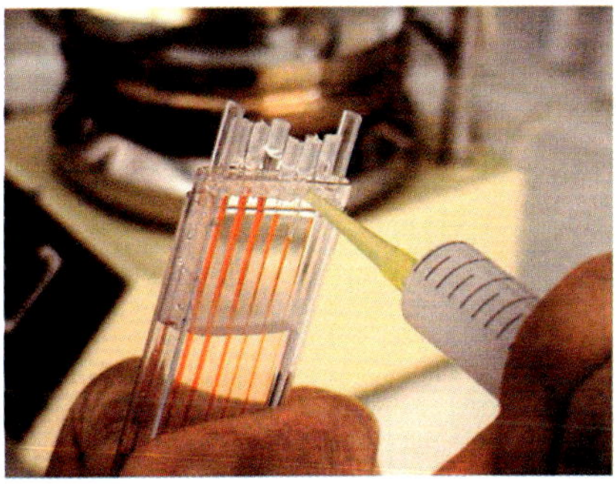

Plate 5.2. Granada Crystallization Box (GCB).

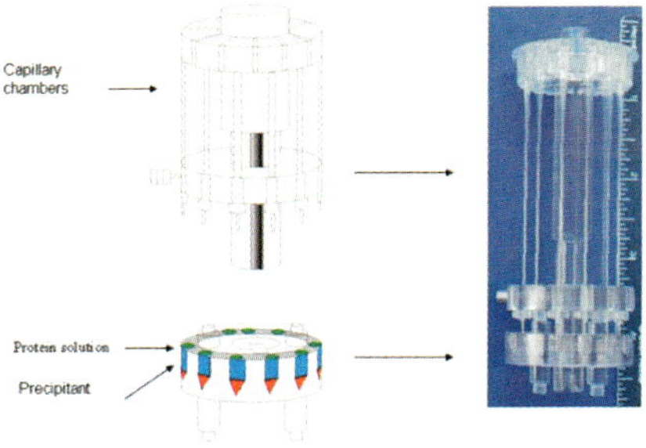

Plate 5.3. Crystallization Cassette.

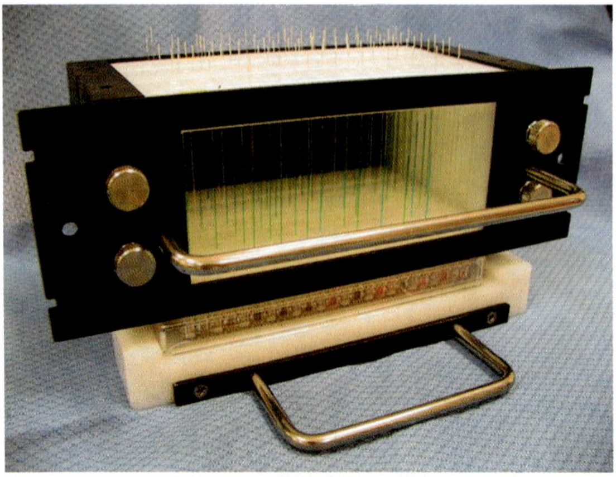

Plate 5.4. 96-capillary format counter-diffusion. Courtesy of Larry DeLucas.

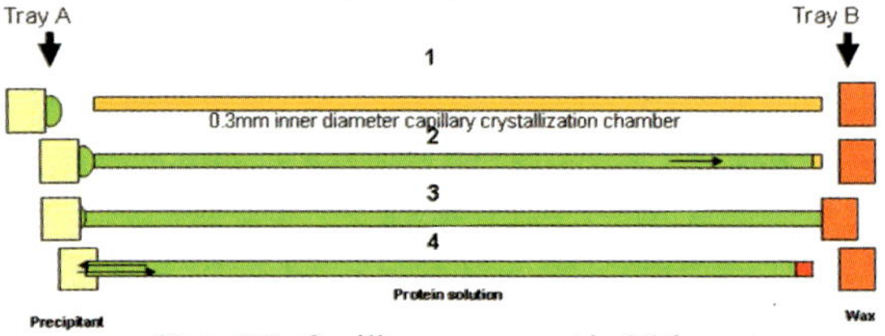

Plate 5.5. Capillary movement in 96 format.

Plate 5.6. Section of capillary tube containing crystals of interest mounted on a goniometer head.

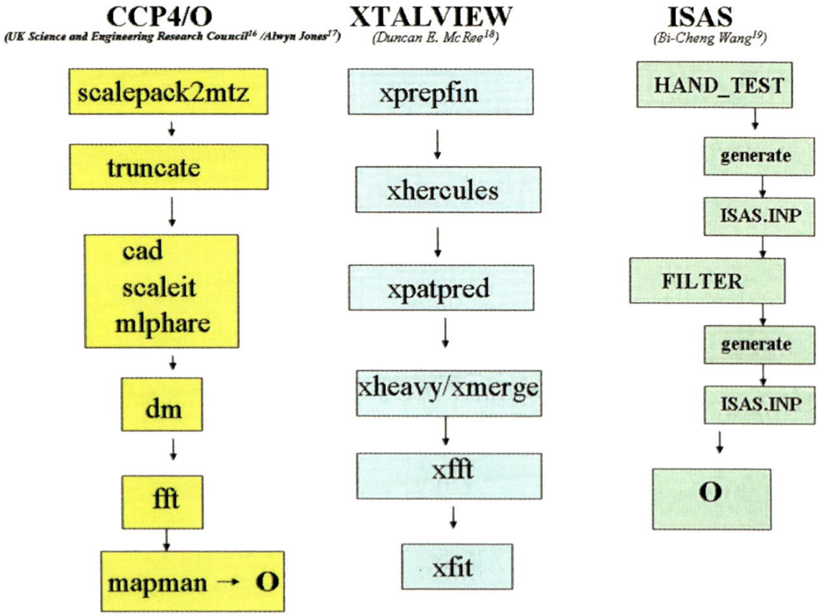

Plate 5.7. Conventional software packages for crystallographic software determination.

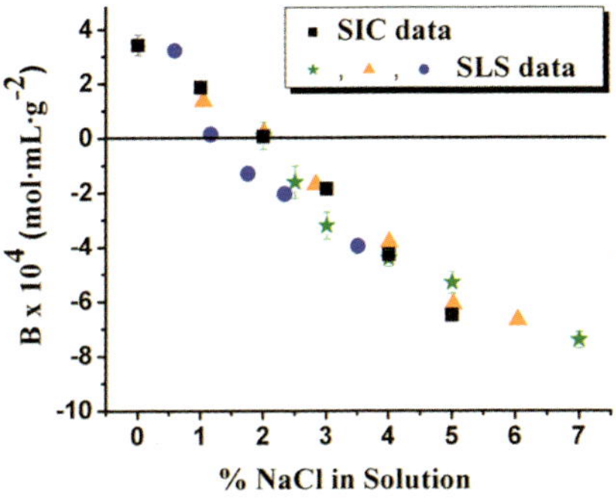

Plate 6.1. Correlation between B values measured by PEEK/SIC and SLS.

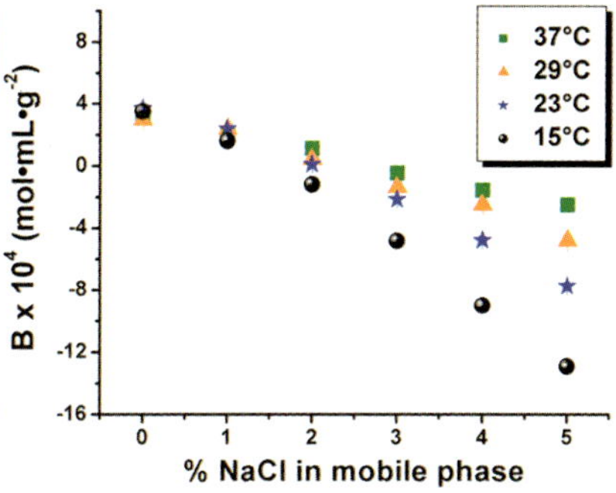

Plate 6.2. B as a function of NaCl concentration and temperature for lysozyme. All values were obtained by PEEK/SIC, and the total analysis time was approximately 18 h for all 24 data points performed in quadruplicate.

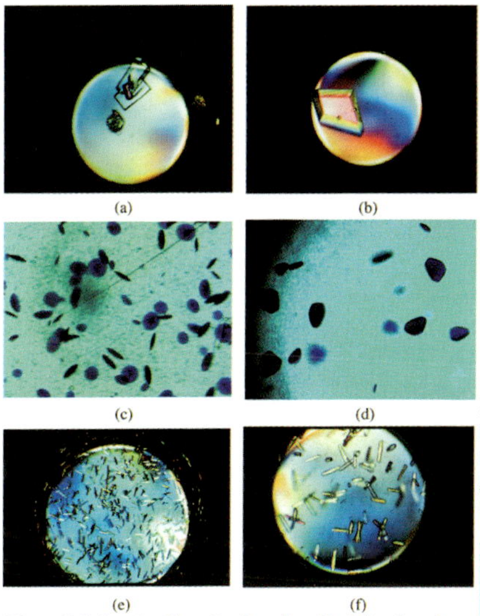

Plate 7.1 (top). Standard and gelled 2-μL microbatch drops (from reference[13] with permission): (**a**) Trypsin crystals in a standard drop. Scale: 1 cm = 0.25 mm; (**b**) Trypsin crystal in a gelled (TMOS) drop under otherwise similar conditions as in (**a**). 1 cm = 0.23 mm; (**c**) C-phycocyanin crystals in a standard drop. 1cm = 0.2 mm; (**d**) C-phycocyanin crystals in a gelled (TMOS) drop under otherwise similar conditions as (**c**). 1 cm = 0.2 mm; (**e**) Concanavalin A crystals grown in sub-optimal conditions in a standard drop. 1 cm = 0.28 mm; (**f**) Concanavalin A crystals grown in TMOS gel under the same conditions as (**e**). 1 cm = 0.28 mm.

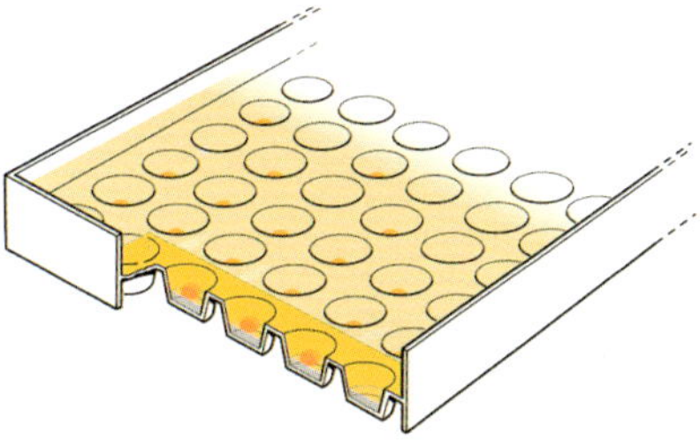

Plate 7.2. Containerless crystallization setup (from reference[13] with permission). Schematic diagram of crystallization drops positioned between silicone grease (in *gray*) and low-density oil (*yellow*) inside a microbatch plate.

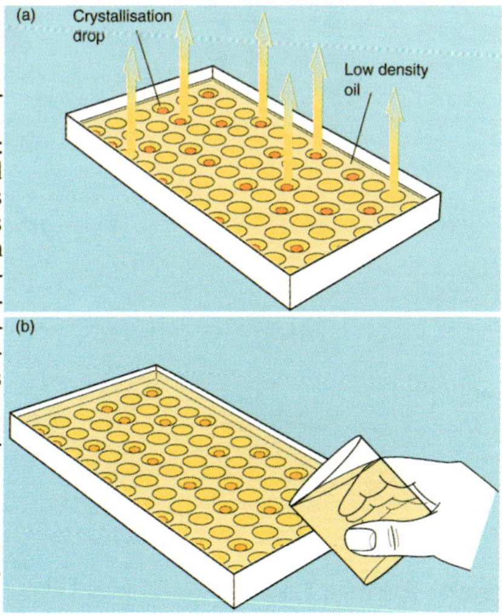

Plate 7.3 (right). Schematic display of controlled evaporation. (from reference[13] with permission). (**a**) Microbatch drops under a thin layer of oil, which allows their concentration (symbolized by the arrows). (**b**) Arrest of evaporation/concentration by topping up the oil to produce a thicker layer above the drops. From then onwards, the experiment follows its "conventional" route.

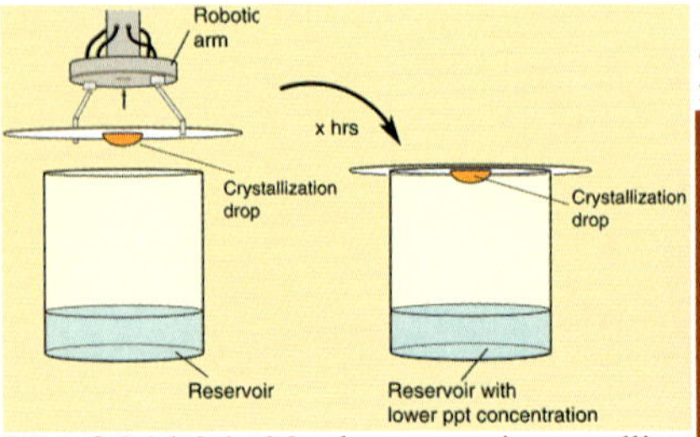

Plate 7.4 (left). Automatic transfer of hanging drops from nucleation to growth conditions. (from reference[13] with permission).

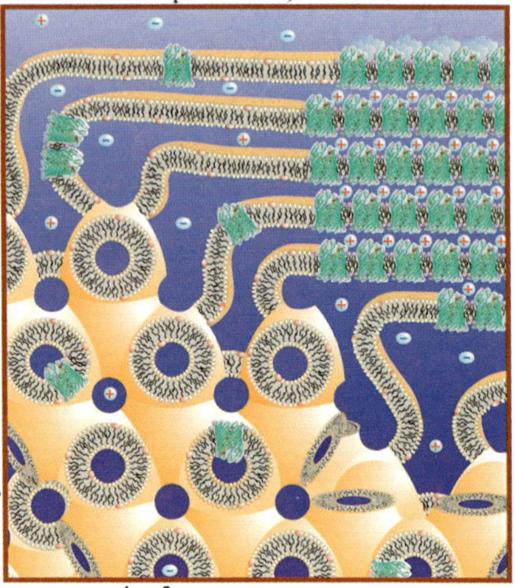

Plate 8.1 (right). Membrane protein crystallization in tailored lipidic mesophases. Cartoon representation of the events proposed to take place during the crystallization of an integral membrane protein from the lipidic cubic mesophase.[1] As much as possible, the dimensions of the lipid (*light brown oval with tail*), detergent (*pink oval with tail*), native membrane lipid (*purple oval with tails*), protein (*blue*; outer membrane Vitamin B_{12} transporter, BtuB; PDB code 1NQE), bilayer and aqueous channels (*purple*) have been drawn to scale. The lipid bilayer is approximately 40 Å thick.

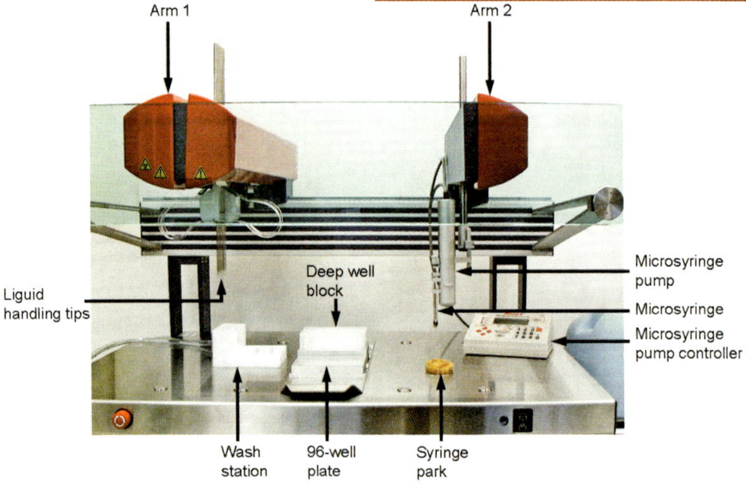

Plate 8.2. The *in-meso* crystallization robot.

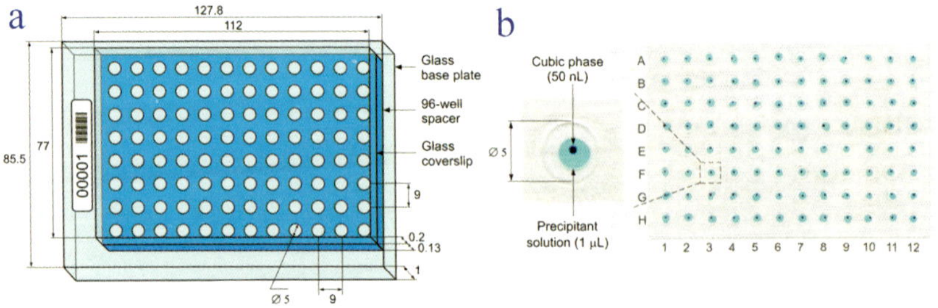

Plate 8.3. The 96-well glass sandwich plate used for crystallization with the in-meso robot. (**a**) Schematic diagram of the plate. Dimensions are given in millimeters. (**b**) A fully loaded and sealed 96-well crystallization plate.

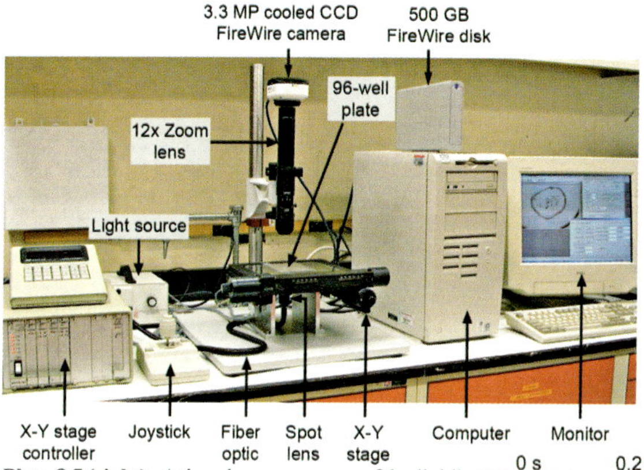

Plate 8.4 (left). The imaging robot for automatic screening of 96-well glass sandwich crystallization plates.

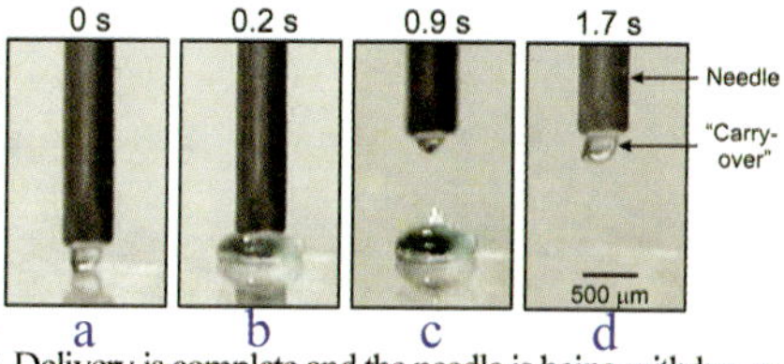

Plate 8.5 (right). A time-lapse sequence of the lipidic cubic phase delivery process. 100 nL of the cubic phase is being delivered on a glass surface. Elapsed time is indicated above each panel. (**a**) "Carry-over" from the previous well is seen at the needle tip just before a fresh bolus is dispensed into the new well. (**b**) The bolus of cubic phase is being dispensed and is filling the space between the needle tip and the glass plate surface. The distance between the needle tip and the plate surface is 250 µm. (**c**) Delivery is complete and the needle is being withdrawn from the plate. The dispensed cubic phase bolus takes on the shape of a flattened teardrop. Because the cubic phase is highly viscous and does not flow, the teardrop shape is stable in time. (**d**) The syringe and needle have moved on to the next well and what is referred to as "carry-over" is seen developing at the needle tip. The latter is explained in more detail in the text (**Section 8.6.1**).

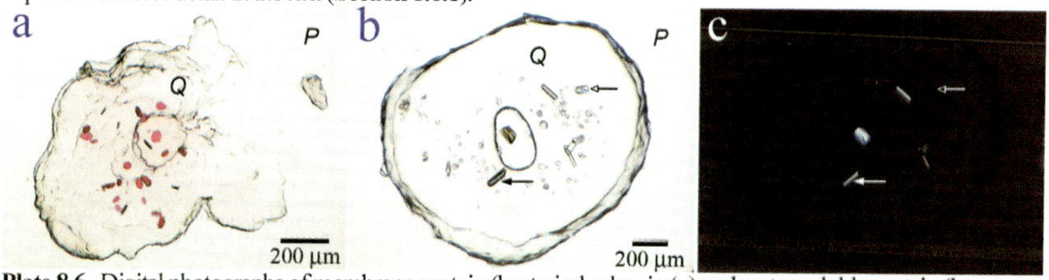

Plate 8.6. Digital photographs of membrane protein (bacteriorhodopsin (**a**) and water-soluble protein (lysozyme (**b** and **c**)) crystals grown using the *in-meso* robot and recorded using the imaging robot. All crystallizations were performed in the monoolein cubic mesophase at 20°C. The bolus of mesophase (50 nL in **a** and 200 nL in **b**, labeled Q) is in the center of the image and is surrounded by precipitant solution (labeled P). The protein crystals in **a** are reddish-purple in color while those in **b** and **c** are colorless. Images **a** and **b** were recorded with polarized light. The image in **c** is the same as in b except that it was recorded between crossed-polarizers. The open headed arrows in **b** and **c** point to a nonbirefringent bubble. The solid headed arrows in **b** and **c** point to a birefringent crystal.

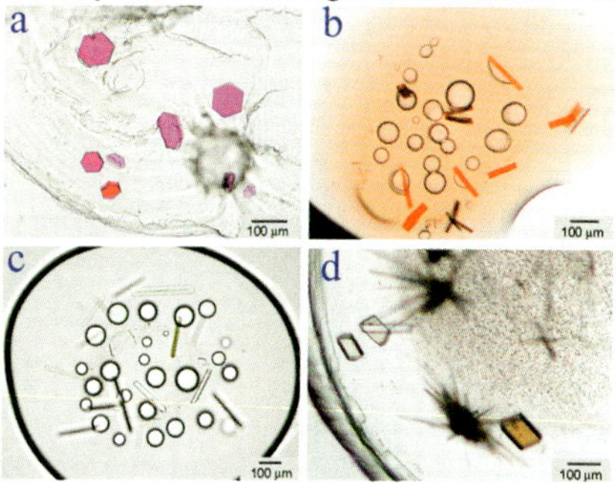

Plate 8.7 (left). Protein crystals grown *in-meso* using the robot. (**a**) Crystals of bacteriorhodopsin. (**b**) Crystals of LH2. (**c**) Crystals of BtuB. (**d**) Crystals of lysozyme.